To Baloo,

with best wishes,

Maurice

Competitiveness and the State

Competitiveness and the State

Government and business in twentieth-century Britain

edited by
Geoffrey Jones *and* Maurice Kirby

Manchester University Press
Manchester and New York
Distributed exclusively in the USA and Canada by St. Martin's Press

Published by Manchester University Press
Oxford Road, Manchester M13 9PL, UK
and Room 400, 175 Fifth Avenue,
New York, NY 10010, USA

Distributed exclusively in the USA and Canada
by St. Martin's Press, Inc.,
175 Fifth Avenue, New York, NY 10010, USA

British Library cataloguing in publication data
Jones, Geoffrey *1952–*
Competitiveness and the state: government and business in twentieth-century Britain.
1. Great Britain. Business enterprise
I. Title II. Kirby, M. W. (Maurice W.)
338.480941

Library of Congress cataloging in publication data
Competitiveness and the state: government and business in twentieth century Britain / edited by Geoffrey Jones and Maurice Kirby.
p. cm.
Revised papers from a meeting held at the University of Reading, 28 October 1989.
Includes index.
1. Industry and state—Great Britain—History—20th century.
2. Trade regulation—Great Britain—History—20th century.
3. Competition—Great Britain—History—20th Century. I. Jones. Geoffrey. II. Kirby, M. W.
HD3616.072C58 1991
338.S41'009'04—dc20 90–13526

ISBN 0 7190 3276 8 *hardback*

Typeset in Great Britain
by J&L Composition Ltd, Filey, North Yorkshire

Printed in Great Britain
by Billings Ltd, Worcester

Contents

List of tables

List of contributors

Dr Jonathan Brown is Business Records Officer in the Institute of Agricultural History, University of Reading. His main research interest is in rural trade and farming in Britain. He is the author of *Agriculture in England, 1870–1945* (1988).

Dr Martin Chick is Lecturer in Economic and Social History at Edinburgh University and Lecturer in Business History at the Scottish Business School. He has edited one book entitled *Government, Industries and Markets: Aspects of government–industry relations in the UK, Japan, West Germany and the USA since 1945* (1990) and has contributed to journals and other edited volumes. He is currently writing a book on the economic planning of the Attlee Governments.

Tony Corley has recently retired from the Department of Economics at Reading University. He has published extensively in the field of business history, most notably on the growth of multinational enterprises. His history of Huntley & Palmers was followed by a two-volume history of the Burmah Oil Company. He is at present preparing an account of the early years of the Beecham Group.

Dr Wyn Grant is Reader in Politics and Chair of the Department of Politics and International Studies at the University of Warwick. He has published a number of books on government–industry relations, most recently *Government and Industry: a Comparative Analysis of the US, Canada and the UK* (1989).

Dr Geoffrey Jones is Reader in Business History in the Economics Department of the University of Reading. His previous appointments included Research Fellow at Corpus Christi College, Cambridge and Lecturer in Economic History, London School of Economics. He is the author or editor of eleven books and numerous articles on the history of international business and other topics, including (ed. with P. Hertner) *Multinationals: Theory and History* (1986); (ed.) *British Multinationals: Origins, Management and Performance* (1986); *Banking and Empire in*

Iran (1986); (ed. with R. Davenport-Hines), *British Business in Asia since 1860* (1989); and (ed.) *Banks as Multinationals* (1990). Dr Jones is co-editor of the journal *Business History* and is currently completing a history of British multinational banking. He is Vice-President of the Association of Business Historians.

Dr Maurice Kirby is Reader in Economic History in the Economics Department of Lancaster University. He previously lectured in economics and economic history at the universities of Nottingham, Heriot-Watt, and Stirling. His publications include *The British Coalmining Industry, 1870–1946: A Political and Economic History* (1977), *The Decline of British Economic Power since 1870* (1981), and *Men of Business and Politics* (1984). His main research interest is the history of early railway company management in Britain and he is currently writing an economic history of trade and industry in the UK since 1870. Dr Kirby is a Fellow of The Royal Historical Society.

Dr Helen Mercer lectures in Economic History at the School of Business and Economic Studies, University of Leeds. Her current interests are the history of competition policy and regulation in Britain, British trade associations, and the 1945–51 Labour governments and the private sector. Her publications include H. Mercer, N. Rollings and J. Tomlinson (eds.), *The 1945–51 Labour governments and private industry* (1991).

Dr Mary Rose is Lecturer in Economic History at Lancaster University. Her main research interests lie in the field of business and textile history, on which she has published a number of articles and a book entitled *The Gregs of Quarry Bank Mill: The Rise and Decline of a family firm 1750–1914* (1986). She is currently working on a comparative study of business development in the British and American cotton industries. Dr Rose is a member of the Council of the Association of Business Historians.

Dr Jim Tomlinson is Reader in the Economics Department, Brunel University and Fellow of the Business History Unit, LSE. He has recently published *Public Policy and the Economy Since 1900* (1990) and *Hayek and the Market* (1990), and is currently working on the economic policies of the Attlee Government.

Oliver Westall is Lecturer in Economics at Lancaster University. He is principally interested in competition and collusion in British business, particulary in insurance, and is the editor of *The Historian and the Business of Insurance* (1984). He is also interested in the interaction between competitive structure and corporate culture, and the impact of information technology on competition in international insurance markets. Mr Westall is currently completing a business history of a non-tariff insurance company.

Preface

There is now an enormous literature on the theme of 'competitiveness', much of it originating in the current preoccupations within the United States about the relative decline of the American economy, and the extraordinary growth of that of Japan. The publication of Michael Porter's new volume on *The Competitive Advantage of Nations* in 1990 has raised the level and depth of the discussion about why the competitiveness of nations varies.

Most commentators are agreed that government policies, whether in the form of regulatory frameworks or direct measures of intervention, have played an important role in determining the competitiveness of individual economies. In that light the essays in this book examine the ways in which governments have had an impact, for better or worse, on the competitiveness of the British economy in the twentieth century, in many ways the weakest of the economies examined by Porter. The approach is historical and the accent in chronological terms is on the years from 1930 to 1970, a period which witnessed a remarkable transition in British governmental attitudes to the issues of productive efficiency and competitiveness. It cannot be claimed that the present volume has achieved complete chronological or sectoral coverage. The aim rather has been to select for analysis critical periods in the evolution of official thinking on competitiveness together with a number of case studies which help to elucidate the most relevant themes.

This book grew out of the close links between the business historians working in the Economics Departments of the Universities of Lancaster and Reading. Over the last decades business history has ceased to consist of company case studies, and has become more conceptual and thematic in its approach, although retaining a commitment to high quality historical research. The presence of business historians working within major British economics departments is one sign of this development. This volume is a modest contribution to this trend. Each author was specially commissioned to write on his or her subject, and to focus on the theme of the 'State and

Competitiveness'. Although the approaches — and conclusions — vary, we believe that our contributors have fulfilled their remit. Three of the contributors (Kirby, Rose and Westall) are from Lancaster, and three from Reading (Brown, Corley and Jones). We are grateful to Martin Chick (Edinburgh), Wyn Grant (Warwick), Helen Mercer (Leeds) and Jim Tomlinson (Brunel) for lending their support to this project and widening the range of expertise available to examine this theme.

Initial versions of most of these chapters were discussed at a conference held at the University of Reading on 28 October 1989. We would like to thank the Nuffield Foundation and the Economics Department of the Unversity of Reading for providing the financial assistance which made this meeting possible.

Geoffrey Jones
Maurice Kirby
June 1990

1 *Geoffrey Jones and Maurice Kirby*

Competitiveness and the State in international perspective

1 Governments and competitiveness

This book is concerned with the impact of government policy during the twentieth century on the competitiveness of the British economy. Competitiveness is defined in both macro- and micro-economic terms. At the macro level the main focus is on competitiveness in the international setting and the role of the state in enhancing Britain's comparative economic performance. It is conventional in this respect for economists to refer to price and non-price competitiveness, in the former case by focusing on money, wages, productivity and the exchange rate, and in the latter by analysing such issues as product specification and reliability in performance. The relevant essays in this volume by Kirby and Rose, Tomlinson and Grant examine the attitudes of government to the issue of productive efficiency defined in its broadest sense. The remaining essays deal with micro-economic performance at the industry and institutional level.

Two themes can be identified. The first concerns the maintenance of 'workable competition' as 'the kind of market pressure which must be exerted to penalise laggards and to reward the enterprising, and in this way to promote economic progress'.[1] The second theme introduces a necessary qualification in that the state has combined competition policies with a desire to maintain security of supply of strategic commodities such as oil, together with a regulatory framework, notably in the financial services sector, with the aim of sustaining commercial stability and confidence. Clearly, the macro- and micro-economic division is not watertight in the sense that measures to enhance competitiveness at the micro level can have a favourable impact on overall economic performance. It is certainly not the intention of the contributors of this volume to erect an artificial dichotomy. The unifying link is the developing concern of British governments in an era of comparative economic decline to improve competitive performance in both world and home markets.

Orthodox economic theory traditionally gave little attention to the role, if any, governments can play in enhancing the competitiveness of their economies. The need for governments to intervene in case of 'market failure' was recognised in conventional welfare economics, but in the context of an overall assumption that the optimum efficient use of resources — known as Pareto-efficiency — was achieved by perfectly competitive markets. There was little room in such a world for governments to make a *positive* contribution to national competitiveness. Undoubtedly, one reason for this situation was the dominance of modern economics by Anglo-American scholars, who drew on, and reinforced, the liberal laissez-faire traditions of their societies.

In contrast, economic historians, especially those who looked beyond the narrow confines of Britain and America, knew that governments had long sought after — and sometimes succeeded in — enhancing the competitiveness of their economies, even though they would not have used that particular term. In the early 1950s Alexander Gerschenkron showed how backward nations in the nineteenth century, such as Tsarist Russia, used the government as a direct development tool and as a substitute for deficiencies in the supply of modern entrepreneurship.[2] Russia, indeed, had a long tradition of governmental attempts to modernise its economy, stretching from Peter the Great in the seventeenth century through to the policies of Count Witte, the Finance Minister of the 1890s, and forward to the Soviets.[3] The twentieth-century socialist states saw the attempt by governments to take over all economic functions in order to plan and shape development. Although his approach was flawed, elsewhere governments in market economies continued to follow policies designed to enhance the competitiveness of their economies. The indicative planning seen in France after 1945, and the government intervention in the Newly Industrialised Countries (NICs) of Asia, such as Taiwan and South Korea, are generally regarded as some of the more successful examples of this approach.

However, it is government policies in Japan which have received the most attention as an apparently highly successful example of a government's ability to enhance national competitiveness in a market economy. On some interpretations, the government has been active in each stage of Japan's growth as a major industrial power since the country was 'opened up' to the world in the mid-nineteenth century. In the Meiji period (1868–1912), Japanese governments were active in the promotion of infant industries, in the use of subsidies to promote new industries such as shipbuilding, and in the creation of a modern financial system. Later, the military governments of the 1930s were closely involved, through both military spending and the protection and promotion of Japanese-owned enterprises in industries such as automobiles, with the successful shift of the Japanese economy away from consumer goods towards heavy industries and chemicals.[4]

It was the role of governments in the Japanese 'economic miracle' since the late 1950s which most attracted — or alarmed — Anglo-American writers. Although public ownership in the economy and taxation as a percentage of GNP were both low in this period compared to many Western economies, Japan appeared to operate a highly successful industrial policy, orchestrated by the Ministry of International Trade and Industry (MITI). MITI, through its policy of 'administrative guidance', was observed prioritising key industries, enabling Japanese business to locate the best technology in the world, obstructing the operations of foreign companies in Japan, and generally playing a highly supportive role as Japanese business conquered one industry after another.[5]

The image of Japan Inc., of government and industry working in harmony to promote the economic growth of their country, is well entrenched in popular opinion, although academic specialists have found parts of the story to be unconvincing. In Meiji Japan, after 1868 governments imported, on occasion, inappropriate Western technology for use in agriculture, while defence spending imposed a great burden on the population in the later nineteenth century. Moreover, economic growth at that time was led by the private sector in agriculture and the consumer goods industries.[6] For the contemporary period, detailed studies of individual sectors such as energy and machine tools have revealed that Japanese government policies were sometimes ineffective, ill-advised or simply irrelevant, although it is unclear how 'representative' such single-industry case studies are.[7] It does seem that one of the most positive achievements of MITI and other government agencies was, and is, to create a supportive environment for Japanese business, rather than to influence the performance of companies directly through state funding or orders. However, Japanese industrial policy also drew the boundary of 'market failure' more broadly than in the United States or Britain. Dependence on foreign companies, for example, was regarded as a 'market failure', in marked contrast to the liberal attitude towards inward direct investment seen in Britain and America.[8]

Whatever the complexities of the Japanese situation, American academics have increasingly turned to the example of Japanese industrial policy in their efforts to explain and reverse the relative decline of the United States over the last twenty years. American writers in particular have stressed the negative impact of 'adversarial relationships' between business and government — as well as management and labour — in the United States. In trying to explain the post-1950 decline of the American steel industry and the corresponding growth of Japanese steel production, for example, researchers at Harvard Business School have explicitly drawn attention to the different governmental policies in the two countries. Japan's success in steel in the 1950s and 1960s, Patricia O'Brien has argued, was engineered with economic tools that most Western economies reject out of hand: protectionism, cartels and interfirm agreements. These policies together produced rapid growth, economies of scale and technological efficiency rather than a conspiracy against the consumer. In the US, where explicit interfirm negotiations were prohibited, companies minimized their risk by resolutely avoiding increased investment.[9]

The competition policies of the Japanese and American governments after 1945 appeared to differ sharply. While American anti-trust policies were applied uniformly across industries, the Japanese pursued different policies for different industries, with 'government regulation of each industry ... tailored specifically to that industry'. While American anti-trust policies focused on preventing inter-company co-operation, in the expectation that improved competitive vigour would result, the Japanese achieved 'the remarkable feat of preserving inter-company rivalry within a co-ordinated group of firms'. In the steel industry, for example, MITI operated a set of policies designed to encourage Japanese firms 'to co-ordinate capacity and pricing decisions, while competing on technological efficiency'.[10]

By the 1980s the experience of Japan and other nations, and the comparative industrial decline of the United States and Britain, had led some American and British economists and industrial policy writers to reassess their views on competitiveness and the state. This new school of thought argued that it was incorrect to think of government intervention always as a market distortion. Government micro-economic strategies can improve the comparative advantages of their economies, through industrial and trade policies, spending on education and research and development, through anti-trust policies, and by less direct means, such as promoting a particular economic ethos or business culture. 'For those who favour a let the market decide philosophy', the British economist J. H. Dunning observed in 1988, 'let them ask themselves where Japan, Korea and Taiwan would be today if those countries had adopted such a strategy twenty years ago; or where, for that matter, Germany, France and the USA would be if they had adopted such a stance in the last century'.[11] This view is broadly shared by the contributors to this book. Governments can enhance the competitiveness of market economies, with policies that involve more than clearing away the obstacles preventing the operation of perfectly competitive markets.

2 British competitiveness in the twentieth century

As indicated earlier, the definition of 'competitiveness' is a complex matter and, according to which criterion is adopted, the assessment of an economy's competitive ability can differ. There is an important distinction to be drawn between the competitiveness of an individual business enterprise (which in the case of a multinational will include its performance in a number of countries) and the competitiveness of a national economy. The United Kingdom and the United States have been the world's largest multinational investors throughout the twentieth century; it is debatable how far, and in what ways, the competitiveness of British (or American) owned corporations can be equated with British (or American) competitiveness.[12] Any index of international competitiveness based solely upon the imports and exports of goods may be different from one which takes into account profits and dividends resulting from inward and outward direct investment, and imports and exports of disembodied technology.[13]

Further very popular definitions of competitiveness highlighted recently in the report of the House of Lords Select (Aldington) Committee on Overseas Trade are the share in world trade in manufacturers held by national economies and the degree of import penetration of manufactured goods. Britain's share of world exports of manufacturers has fallen very sharply over the twentieth century. On the eve of the First World War, Britain was the world's largest exporter. Around 40 per cent of these exports were textiles and clothing, followed in importance by iron and steel, coal and machinery. All of these products experienced a severe drop in their overseas markets in the inter-war years. Noting the persistence of this trend after 1945 (see Table 1.1), the Aldington Committee concluded that the UK economy had experienced a serious loss of competitiveness which threatened to undermine the stability of both the economy and the political and social fabric. This was despite the fact that Britain's export composition after 1945 had shifted from low technology products sold to the Third World to more sophisticated goods sold to other industrialised countries, such as chemicals, electrical goods and vehicles.

Table 1.1 *UK share of world trade in manufactures (%)*

1950	25.4	1969	11.2
1954	20.5	1974	8.8
1959	17.7	1979	9.1
1964	14.2	1984	7.6

Source: Table 3.1 in House of Lords, *Report from the Select Committee on Overseas Trade*, Vol. 1 (session 1984–85), p. 23

But this is to ignore the relative openness of the British economy measured by the proportion of GDP exported. Because other economies have moved nearer to the British figure (25–30 per cent for the period 1975–85) the volume of world trade has expanded at the same time as the British share, by the process of simple arithmetic, has declined. Differential rates of economic growth can also contribute to declining trade shares. With a relatively slow rate of growth combined with a stable contribution of exports to GDP the UK has inevitably lost market share. In this case slow growth rather than competitiveness is the causal factor. This, of course, is not to deny that slow growth may itself be the result of a lack of competitiveness. Similarly, a rise in import penetration is not an unequivocal measure of declining competitiveness. In the late nineteenth century, foodstuffs and raw materials were the main imports into Britain. In 1913 only 5 per cent of British imports were manufactured goods. By 1951 manufactured goods comprised 20 per cent of imports, and by 1989 over 62 per cent. In 1983, for the first time since the Industrial Revolution, Britain imported more manufactured goods in value than she exported. In some sectors the extent of import penetration was striking. In 1965 only 5 per cent of British demand for vehicles was met by imports. Ten years later the figure was 33 per cent and by 1990 over 60 per cent. That said, it is important to bear in mind that rising import penetration may be the product, in part, of domestic economic growth and the growing internationalisation of world trade. It is also the case that UK imports in the past contained a high percentage of foodstuffs and raw materials. To the extent that these possessed a relatively low income elasticity of demand, a rise in the proportion of manufactures in total imports was likely to follow in the wake of rising GNP.[14]

Michael Porter is surely right to argue that the key underlying concept in understanding competitiveness is productivity. National competitiveness, according to Porter, should be judged by an economy's capacity to raise living standards by continuously improving productivity in the output of ever more sophisticated goods and services.[15] By this standard, Porter has no doubt that Britain has had, and continues to have, a problem of competitiveness.

Worries about definitions have rarely perturbed the generations of economic historians and others who have contributed to the voluminous literature on Britain's industrial decline in the twentieth century. The origins of Britain's problems have been firmly placed as far back as the late Victorian era, when the economy's 'faltering technological ability' led to the retarded development of certain 'new' industries such as electrical engineering, telephone communications and dyestuffs.[16] By the 1970s and early 1980s, Britain's international competitiveness had apparently declined so much that some academic observers, in addition to members of the House of Lords, were sunk into the deepest despair. 'Within the span of half a lifetime', Sidney Pollard wrote in 1982, 'Britain has descended from the most prosperous major state of Europe to the Western European slum ... She has become the proverbial failure.' By around 2010, Pollard predicted, Britain would be 'the poorest country in Europe, with the possible exception of Albania.'[17]

The 'decline' literature is frequently prone to exaggeration. In the 1980s the absolute size of the British economy remained one of the largest in the world, behind only the United States, Japan, the Federal Republic of

Germany, Italy and France. As Table 1.2 shows, up to 1950 the United Kingdom had a larger real GDP than France, Germany or Japan, though it was dwarfed by the United States from the late nineteenth century. However, by 1973 Germany and Japan had passed Britain, and by 1984 France's GDP was also larger than that of Britain.

Table 1.2 *Real GDP at 1984 purchasing power parity, 1870–1984 ($ billion)*

	UK	*France*	*Germany*	*Japan*	*USA*
1870	77.95	59.27	33.98	19.28	78.61
1913	174.78	119.99	111.75	54.76	454.53
1950	281.04	173.49	179.92	124.34	1,257.86
1973	556.60	547.98	675.49	976.50	2,911.78
1984	625.20	694.70	811.60	1,468.40	3,746.50

Source: Angus Maddison, 'Growth and slowdown in advanced capitalist economies: techniques of quantitative assessment', *Journal of Economic Literature*, XXV, 1987, p. 682

The scale of Britain's economic misfortune, therefore, is less than writers such as Pollard imply. The main outline of events is clear. Britain had lost its position as the world's largest industrial economy to the United States by the beginning of the twentieth century. British manufacturing industry remained a giant in the world economy, but was notable for its low value added, low skill, highly specialised orientation. In the inter-war years Britain's industrial, and underlying productivity, performance improved, as industries such as chemicals, electricity generation and motor cars 'caught up' with developments in the United States and Germany. Post-1945 Britain's industrial performance improved further, measured against the economy's previous record. Unfortunately, Britain failed to experience an 'economic miracle' of the kind that transformed the economies of Continental Europe and Japan in the 1950s and 1960s.[18]

Table 1.3 gives the GDP growth rates of Britain, the United States, France, Germany and Japan between 1870 and 1984. Britain kept up with two other western European economies in the inter-war years (as it was to

Table 1.3 *Phases of GDP growth, 1870–1984 (average annual compound growth rates)*

	1870–1913	*1913–50*	*1950–73*	*1973–84*
UK	1.9	1.3	3.0	1.1
France	1.7	1.1	5.1	2.2
Germany	2.8	1.3	5.9	1.7
Japan	2.5	2.2	9.4	3.8
USA	4.2	2.8	3.7	2.3

Source: Maddison, p. 650

do again between 1986 and 1988), but the overall picture was of an economy growing more slowly than its competitors. The differentials with the western European countries and Japan were particularly large in the era of high growth between 1950 and 1973.

Slow productivity growth — Porter's ultimate test of national competitiveness — was at the core of Britain's relative economic decline. As Tables 1.4 and 1.5 demonstrate, UK labour and capital productivity growth tended to lag behind competitor countries in most periods. British productivity performance was particularly poor in the 1950s and 1960s, although compared to its past record labour productivity at least was growing faster than ever before. As a result of these new growth rates, absolute levels of productivity in Britain were very much lower than those of competitor countries by the late 1980s: perhaps 30 per cent less than in the United States and 25 per cent less than the European Community average.

Table 1.4 *Growth in labour productivity (GDP per hour worked, 1870–1984) (average annual compound growth rates)*

	1870–1913	*1913–50*	*1950–73*	*1973–84*
UK	1.2	1.6	3.2	2.8
France	1.7	2.0	5.1	3.4
Germany	1.9	1.0	6.0	3.0
Japan	1.8	1.7	7.7	3.2
USA	2.0	2.4	2.5	1.0

Source: Maddison, p. 656

Table 1.5 *Growth in capital productivity, 1913–84*

	1913–50	*1950–73*	*1973–84*
UK	0.13	−0.26	−1.45
France	0.12	1.50	−1.82
Germany	0.56	0.57	−1.71
Japan	0.69	1.39	−3.41
USA	0.96	0.34	−0.47

Source: Maddison, p. 656

Porter's work confirms that Britain's competitiveness has continued to decline in the 1970s and 1980s, despite some productivity improvements in the 1980s. Although by 1990 Britain retained significant competitive positions in a number of industries, such as chemicals, pharmaceuticals, financial services and some consumer goods, such as biscuits, 'far more competitive industries in Britain (since 1978) have lost world export share than have gained it', and losses have been concentrated in 'sophisticated industries'. Britain's economy has continued, in the words of the sub-heading in Porter's book, to 'slide'.[19]

Explanations of the causes of the inferior British economic performance are legion. Modern business historians have stressed inadequate business organisation and management. Alfred D. Chandler, the doyen of business historians, has emphasised the deficiencies of British family firms which lingered on decades after more professional management structures developed elsewhere. For him, British 'personal capitalism' was the epitome of failure compared to the American system of 'competitive managerial capitalism' and German 'co-operative managerial capitalism'.[20] The British industrial relations system, an alleged 'anti-industry' bias in the City and British banks, low levels of education and training, and a socio-cultural environment antipathetic to the 'industrial spirit' are other frequent culprits.[21]

Porter argues that the mechanisms for creating national competitive advantage form a 'diamond' representing the interaction of four determinants. These are the ability of economies and industries to create factor advantages by upgrading investment, research and skills; the quality of home market demand; the proximity of competent suppliers and 'clustering' by closely-related industries; and the intensity of local competition, which for Porter is probably *the* single most important factor. The United Kingdom performs poorly in all four areas in his analysis. The relative decline of British living standards after the Second World War made the market unattractive, and resulted in consumers 'resigned to poor service or substandard quality'. An extremely elitist and inadequate educational system contributed towards low skill levels. Most importantly, Britain had a 'management culture that works against innovation and change', where domestic rivalry between firms was weak and 'competition is viewed as distasteful and even vulgar'.[22]

It is clear that there is no monocausal explanation of Britain's faltering competitiveness, and that some or all of these explanations form part of an overall explanation. The focus of the following chapters, however, is the extent to which the British government can be held responsible.

3 The British government and competitiveness

The macro-economic policies of the British government in the twentieth century have received formidable attention and, although much of the discussion has not been couched in the language of 'competitiveness', the impact on Britain's economy has been clear. The most celebrated issue to have been discussed was the alleged overvaluation of sterling by around 10 per cent following Britain's return to the gold standard in 1925, which had an adverse impact on the price competitiveness of British exports. The subsequent depreciation of sterling in 1931, conversely, had at least the potential to increase the competitiveness of British exports. However, the generally accepted opinion appears to be that these exchange rate policies had comparatively little impact on export performance.[23] After 1945, the British obsession with the defence of sterling, and the use of interest-rate manipulation and credit controls in the 'stop–go' mechanism, have been extensively analysed.[24] The British government's determination to promote home ownership through tax concessions probably also had negative effects on economic performance. Certainly such policies contrasted with those of Japan, where widespread home ownership was not encouraged, thus contributing both to high savings ratios and the diversion of funds for investment in industry.[25]

Yet the general conclusion appears that though British macro-economic policies were not ideal, they made a bad situation worse rather than serving as a fundamental cause for declining competitiveness. 'Stop–go' policies were pursued by governments in post-war Continental Europe as well as Japan and obviously did not damage their competitiveness. Perhaps the most worrying feature of British macro policies is that they may heve been used as a substitute for more badly needed micro policies which, in political and social terms, were too painful to pursue. Keith Middlemas, for example, has recently shown from the 1960s and early 1970s how successive governments preferred to raise demand as a means to achieve higher growth rates, rather than attempt policies to improve Britain's dismal productivity record.[26]

The micro-economic context of British government policy has received less attention. The most influential work has come from the United States and by scholars influenced by the 'industrial policy' debate. In 1986 Elbaum and Lazonick's *Decline of the British Economy* offered a new institutional interpretation of Britain's relative industrial decline over the last hundred years, and implicated government policies in that decline.[27] Elbaum and Lazonick took as their central point of reference Alfred Chandler's model whereby the institutions of competitive capitalism — the 'invisible hand' of the market — were replaced by the visible hand of corporate bureaucratic management. It was the latter which underpinned the successful development of the American and German economies after 1900, and it was Britain's failure to emulate this experience which largely accounted for the country's industrial retardation.

Elbaum and Lazonick blamed the retention of Britain's atomistic economic organisation of the nineteenth century for placing institutional constraints on economic progress. Given the inadequacies of British management, Elbaum and Lazonick argued that the government should have acted as a 'visible hand' to reorganise and modernise industry. 'State policy is implicated in British decline', they suggested, 'by virtue of its failure to intervene in the economy more decisively in order to take corrective measures or to alleviate the social cost of industrial obsolescence and structural change'.[28] The half-hearted attempts by the government and the Bank of England to modernise Britain's staple industries in the inter-war years provided the core empirical evidence to support this case. Steve Tolliday subsequently published a large-scale study of the inter-war steel industry which analysed the profusion of conflicting interests which worked to block the necessary large-scale amalgamations and investment in the industry. The Bank of England and government singularly failed to act as 'visible hands' to break the deadlock, largely for reasons of political and financial expediency.[29]

The Elbaum and Lazonick thesis can be criticised on several grounds. In the first instance it is salutary to remember that economic growth in the UK was sustained throughout the period covered by the thesis: decline, as noted already, has been relative in international terms, whereas the explanation proferred by Elbaum and Lazonick — its determinism and all pervasiveness — might be regarded as consistent with absolute decline. The historical growth record, therefore, suggests that institutional structures, rather than presenting absolute or insurmountable barriers to economic growth have acted as a retardative influence. Furthermore, even if the Elbaum and Lazonick thesis presents a convincing explanation of decline in its own terms, it is nevertheless based on a limited number of industrial case studies. Institutional impediments to growth may have loomed large in cotton textiles, iron and steel, shipbuilding and motor vehicles, but as everyone familiar with the historiography of decline is aware the historic performance of British business is not one of unrelieved gloom, especially if sectors other than manufacturing are taken into account.[30] There must also be doubts about the universal applicability of the 'Harvard Business School-approved diversification strategy and multidivisional structure'. The experience of the Lancashire Cotton Corporation in the early years of its existence after 1929, when an inappropriate divisional structure was soon replaced by 'a centralised, functionally departmentalised management structure', underlines this point,[31] as does a recent study of the inter-war coal industry.[32] The latter, based on a painstaking analysis of individual companies, concluded that the gains to be derived from multidivisional structures in an industry which provided unfruitful ground for continuous-process, assembly line technology, were marginal. The same study also demonstrated that some firms in particular coalfields proved capable of surmounting institutional constraints, and to the extent that similarly placed firms failed to follow suit the thesis of entrepreneurial failure, much derided by Elbaum and Lazonick, comes back into play almost by default.

One effect of these reservations is to highlight the fact that the Elbaum and Lazonick thesis offers a predominantly supply-side explanation of Britain's industrial decline. Yet there is a long sequence of studies which has drawn specific attention to the role of demand-side forces in retarding the introducuion of mass production technology and large-scale corporate enterprise.[33] The contrast has invariably been with the USA in the latter half of the nineteenth century where the internal market was growing rapidly on the basis of homogeneous consumer tastes leading to standardised patterns of consumption. In Britain, class distinctions were more finely drawn and the resulting product differentiation helped to divide and sectionalise the market in ways which proved inimical to the technology of mass production. This is not to suggest that the nature of demand provides a more complete explanation of Britain's industrial retardation. It is instead to suggest the need for an additional dimension to the supply analyses of Elbaum and Lazonick and others.

Nevertheless, in the context of the present volume the Elbaum and Lazonick thesis has succeeded in opening up for discussion the question of whether British government micro policies are implicated in Britain's relative decline and, if so, how and why. The essays in this book seek to make a contribution in this specific area.

As is clear from the essay by Kirby and Rose, British governments were aware of problems of declining competitiveness even before the First World War, and the Lloyd George government, via the Minisry of Reconstruction, analysed in detail the case for state-sponsored programmes of industrial modernisation in peacetime. The post-1945 Labour government was, as Tomlinson shows, very much concerned with the question of raising British productivity performance. Nationalisation was, in part, a strategy to improve the competitiveness of a group of industries whose performance in the inter-war years had been dismal. For private industry there was a concern to regulate monopolies and cartels in order to increase industrial efficiency. Post-1945 agricultural policy and policies to control the 'abuses' arising from restrictive business practices to improve competitiveness saw much talk about the need for 'efficiency'.

Some of the government intervention in areas of 'market failure' was relatively successful. Inter-war governments were effective in reorganising the chemical industry by encouraging the formation of ICI in 1926, which laid the basis for a highly successful chemical industry. The establishment of the National Grid and related changes transformed the productivity of British electricity generation. In the 1930s government intervention created a modern and successful passenger transport system for London.[34] In other cases, such as shipbuilding, the government proved better at reducing capacity than achieving modernisation.[35] After the Second World War, and especially — as Grant shows — from the 'Brighton Revolution' of 1960, British governments were trying various means to raise the competitiveness of British industry. The Ministry of Technology, the Industrial Reorganisation Corporation and the National Enterprise Board were among the institutional innovations of the 1960s and 1970s which were designed to assist improvements in competitiveness. Some of these agencies were moderately successful, and are now attracting the attention of proponents of an American industry policy. The National Enterprise Board, for example, was successful as a venture capitalist supporting firms in high growth markets, though rather less good at actually establishing new firms.[36]

It is less obvious that governments were able to diagnose correctly the reasons for Britain's problems. It is most unlikely, for example, that the key to improving British industrial efficiency in the early 1900s lay merely in promoting co-operation between employers and employees. Even when accurate diagnosis was made, as reflected in the long established view that the relative decline of the economy was in some way related to deficiencies in the provision of scientific, technical and managerial education a combination of factors conspired to pre-empt a sustained and vigorous state response. In the twentieth century these factors have ranged from complacency and the desire for economy in public expenditure, to the low status of engineers and businessmen in society and the deeply rooted cult of practical experience.[37] Again diagnosis and policy were far from identical. The 1945–51 Labour government was correct in believing that British productivity needed improving, but the government's actual policies, especially as a result of the overwhelming concern with the balance of payments and export expansion, sought to increase productivity as a means of obtaining increased output from more or less given resources, rather than as a means of enhancing competitiveness. Resources were diverted into sectors where immediate balance of payments improvements might be forthcoming or supply contraints lifted — textiles, agriculture, coal — rather than where long-run growth might be anticipated.

Two aspects of government policy emerge particularly clearly. First, governments have been extremely reluctant to move against vested interest groups in the British economy. The blocking power of industrialists is a major theme in Mercer's examination of the reasons for the stunted development of Britain's anti-trust legislation. Chick shows the government's subordination to the power of the electricity managers. In the financial sector, although there were widespread complaints about the failure of banks to support British industry and about various 'gaps' in the provision of funds, the government was most reluctant to challenge the City and the banks, and official attempts to do so were far from vigorous. During the First World War official promotion and support of 'trade banks' such as the British Trade Corporation faltered when faced by opposition in the City. After the Second World War the Industrial and Commercial Finance Corporation, a body created by the government to fill a perceived need for external funds from small and medium-sized firms, got off to a very slow start because of opposition from the large clearing banks, which successive governments were unwilling to overcome.[38] Twentieth-century Britain comprised a myriad of stable and well-organised special interest groups, which British governments felt neither the desire nor the energy to challenge. Perhaps the major exception to this generalisation was the willingness of Conservative governments in the inter-war years and the 1980s to attack and curb the powers of trade unions. Unfortunately, such policies were related more to political considerations than any concern for improving competitiveness.

The sclerotic tendencies of long established institutions provides the focal point of Olson's study of the rise and decline of nations.[39] In the British case in the twentieth century he notes that victory in two world wars allowed interest groups to consolidate their position in marked contrast to continental Europe and Japan where defeat and occupation led to major restructuring of economies and societies. Freed from vested interests, countries such as West Germany, France and Japan were able to enjoy rapid rates of economic growth after 1945. The Olson thesis is vulnerable to the argument that accelerated growth was part of a catching up process with the USA in terms of technology and per capita incomes which had been delayed by the dislocating effects of war and economic instability in the 1920s and 1930s. Because the gap between the UK and the USA was no so large the acceleration in growth was not as marked.[40] As far as Britain is concerned it can also be argued that institutional sclerosis would have been greatly alleviated by the development after 1945 of an effective partnership between Whitehall and industry, so that business interest organisations, as with their German and Japanese counterparts, could have become mechanisms for the development and delivery of public policy. One precondition for such a relationship, never fulfilled in Britian, was reform of the chaotic system of business representation. These strictures apart, there is much of value in Olson's work: it certainly lends support to the notion of the 'British disease', and it is a credible argument that Britain's relatively slow convergence to the American per capita income level has been due, at least in part, to institutional rigidities.

A second noteworthy long-term characteristic of British governments was their desire for stability and security. This emerges strikingly in the chapters on agriculture, banking, insurance and oil. Collusive agreements and arrangements were not only permitted, they were encouraged. Sometimes, as in the banking sector in the 1960s, governments wished that collusion should continue even when the industry itself preferred to move towards a more competitive structure. In the case of the nationalised industries, the government took its suspicion of competition to the logical conclusion of taking over and merging together all the firms in an industry.

British industrial policy, therefore, was quite different from that of the United States. As we have seen, Americans have criticised the 'adversarial' relationship between business and government in the United States and seen it as one cause of that country's declining competitivenes. The close relationship between government and business in Japan has been held up as a paradigm. Elbaum and Lazonick tended to assume that British and United

States policies were similar, based on 'the presumption that, if only the government pursued the right fiscal and monetary policies, the operation of the free market would suffice to ensure economic prosperity'.[41] In fact, British governments, at least after the First World War, had little faith in the operation of the free markets a such. Collusion and co-operation were the watchwords. British industrial policy was anything but adversarial. British policy-makers may not have been enthusiastic 'visible hands', but they were scarcely enthusiasts for unregulated markets. Mrs Thatcher's willingness in the first half of the 1980s to privatise British Gas, British Airways and British Telecom as giant private monopolies with very loose regulatory arrangements was very much in the mainstream tradition of British policy.

It is not axiomatic that the restricted competition British governments permitted or encouraged would necessarily inhibit competitiveness, but the work of Michael Porter would certainly suggest that the lack of domestic rivalry in British industry was a considerable influence on Britain's declining competitiveness. The negative consequences can be seen in several directions. On the one hand, promoting collusion slowed down the process of concentration, permitting firms to retain their independence while gaining some monopoly profits. On the other hand, the collusive British market stood, buttressed by the institutions of Imperial Preference, in stark contrast to post-1945 Japan, where fierce competition in many sectors, such as automobiles, served to sharpen the competitive interests of Japanese companies before they penetrated foreign markets. The chapters in this volume highlight some specific losses from the lack of competition in Britain. According to Corley, UK oil prices were higher than elsewhere. Brown shows that by the time Britain entered the European Community in 1973, the heavily subsidised and protected British agriculture had fallen behind the productivity of several other countries. In the banking sector, the collusive and uncompetitive clearing banks lost market share to the building societies, from the inter-war years, and to foreign banks from the 1960s.

British governments recognised the problem of declining British competitiveness. However, partly because of problems of diagnosing the cause, partly out of reluctance to move against vested interest groups, partly through a long-term preference for stability and security, competition was restricted and competitiveness hindered. Given that British business had strong collusive tendencies and a multititude of conservative managements and family firms, government policy can be seen to have worked to exacerbate the comparative disadvantage of the British economy.

Indeed, some sympathy can be felt for Porter's view that, given the inadequacies of British policy-making towards industry, the industries that flourished were the ones that had least to do with the government. Low regulation, for example, enabled British firms in certain service industries such as auctioneering to remain competitive.[42] A passive and welcoming stance towards foreign multinationals helped to make Britain an attractive host country for foreign-owned manufacturing facilities, which have tended to have higher productivity levels than their indigenous counterparts, and be clustered in 'newer' and more dynamic industries.[43]

There are already a number of suggestions for the apparent inadequacies of government policy-making in the twentieth century. Pollard has written of the 'contempt' for industry on the part of a Treasury-dominated civil service elite obsessed with macro-economic issues.[44] Robert Locke has argued that the problem with the British civil service was not necessarily the narrow public school and Oxbridge background of its administrative grade, but the fact that these people were not trained after joining the government. 'From a knowledge point of view', he concludes, 'the British civil servant has not benefited from as good a pre-entry or post-entry education as the French or the Japanese ... To the extent, therefore, that the management–knowledge quotient in the civil service affects British business and industry, they suffer comparatively'.[45] Some stress not the poor quality of British civil servants, but the way in which policy is determined by the institutionalised distribution of power in a society. The British preoccupation with the preservation of sterling, according to Hall and Ingham, was due to the organisation of capital and the state — the international orientation of the City of London and the relative autonomy of the Bank of England.[46] The flawed nature of the British political system, which enables politicians to engage in squalid tinkering with economic policies in order to influence electoral prospects, provides a further set of difficulties. The chapters in this collection make several observations on the indequacies of the government machinery. Corley sees a conflict within government departments between Oxbridge Firsts and the technologists who identified with their clients, while Chick argues that the government was handicapped by a lack of information on the nationalised industries it sought to control.

4 Conclusion

From the turn of the century British politicians and civil servants were well aware of the competitive threat to the British economy. If anything, the problems of the British economy were exaggerated by contemporaries, at least until the post–1945 period when Britain's relative economic decline really did become a serious matter. British governments were less guilty of complacency than of an ability or unwillingness to participate in finding solutions for the underlying British problems of poor productivity and declining competitiveness. They had also to contend with a general electorate eager to enjoy the fruits of economic growth but, in a peculiarly British way, determined to maintain life styles, modes of work, and economic structures whose disappearance was a prerequisite for that growth.

American historians have criticised the adversarial nature of business–government relations in the twentieth century in the United States. Britain had almost the opposite problem. British governments prized stability and consensus above all, preferring stagnation to painful change. They feared provoking the wrath of vested interests. Japanese governments encouraged both co-ordination and competition within Japanese industry. American governments achieved competition in American industry, but little co-ordination. British policies encouraged a situation in British business where there was neither co-ordination nor competition, but rather a collusive and cosy world of low productivity, slow growth and declining competitiveness.

Notes

1 G. C. Allen, *Monopoly and Restrictive Practices*, London, 1968, p. 24.

2 Alexander Gerschenkron, 'Economic backwardness in historical perspective', reprinted in Gerschenkron, *Economic Backwardness in Historical Perspective*, New York, 1962.

3 See the various essays in William L. Blackwell (ed.), *Russian Economic Development from Peter the Great to Stalin*, New York, 1974.

4 W. J. Macpherson, *The Economic Development of Japan c 1868–1941*, London, 1987, pp. 32–43. For a detailed study of the long-term role of the Japanese government in promoting Japanese shipping and shipbuilding, see Tomohei Chida and P. N. Davies, *The Japanese Shipping and Shipbuilding Industries*, London, 1990.

5 Chalmers Johnson, *MITI and the Japanese Miracle; The Growth of Industrial Policy, 1925–1975*, Stanford, 1982; H. Patrick and H. Rosovsky (eds), *Asia's New Giant. How the Japanese Economy Works*, Washington, 1976.

6 W. W. Lockwood, *The Economic Development of Japan*, Oxford, 1955; F. B. Tipton, 'Government policy and economic development in Germany and Japan, a skeptical revaluation', *Journal of Economic History*, 1981.

7 Richard J. Samuels, *The Business of the Japanese State*, Ithaca, 1987; David Friedman, *The Misunderstood Miracle: Industrial Development and Political Change in Japan*, Ithaca, 1989.

8 Daniel I. Okimoto, *Between Miti and the Market: Japanese Industrial Policy and High Technology*, Stanford, 1989.

9 Patricia A. O'Brien, 'Co-ordinating market forces: the anatomy of investment decisions in the Japanese steel industry, 1945–1975', *Business and Economic History*, 1987, p. 209; for a detailed study of the American steel industry from a similar perspective, see Paul A. Tiffany, *The Decline of American Steel: How Management, Labour and Government Went Wrong*, New York, 1988.

10 Thomas K. McCraw and Patricia A. O'Brien, 'Production and distribution: competition policy and industry structure', in Thomas K. McCraw (ed.), *America Versus Japan*, Boston, Mass., 1988, pp. 114–16.

11 J. H. Dunning, *Multinationals, Technology and Competitiveness*, London, 1988, pp. 246–7. For recent American literature on this theme, see J. Zysman and L. Tyson, *American Industry and International Competition*, Ithaca, 1983; B. Scott and B. Lodge, *US Competitiveness and the World Economy*, Boston, 1985.

12 See the article on this theme by Robert Reich in *Harvard Business Review*, January/February 1990.

13 Dunning, *Multinationals*, pp. 59–61.

14 Arthur Francis, 'The concept of competitiveness', in Arthur Francis and P. M. Tharakam (eds), *The Competitiveness of European Industry*, London, 1989.

15 Michael E. Porter, *The Competitive Advantage of Nations*, London, 1990, pp. 6–9.

16 William P. Kennedy, *Industrial Structure, Capital Markets and the Origins of British Economic Decline*, Cambridge, 1987, p. 5.

17 Sidney Pollard, *The Wasting of the British Economy*, Beckenham, 1982, p. 6.

18 The latest overview of the twentieth century British economy is N. Crafts, 'The assessment: British economic growth over the long run', *Oxford Review of Economic Policy*, IV, 1988. See also M. W. Kirby, *The Decline of British Economic Power Since 1870*, London, 1981; C. H. Lee, *The British Economy since 1700: A Macroeconomic Perspective*, Cambridge, 1986; B. W. E. Alford, *British Economic Performance 1945–1975*, London, 1988.

19 Porter, *Competitive Advantage*, pp. 482–96.

20 Alfred D. Chandler, *Scale and Scope*, Cambridge, Mass., 1990.

21 Martin J. Weiner, *English Culture and the Decline of the Industrial Spirit*, Cambridge, 1981. For an alternative 'cultural' explanation, see D. C. Coleman and Christine Macleod, 'Attitudes to new techniques: British businessmen, 1800–1950', *Economic History Review*, 1986, pp. 588–611.

22 Porter, *Competitive Advantage*, pp. 496–504.

23 Derek H. Aldcroft, *The British Economy*, Volume 1, Brighton, 1986.

24 B. W. E. Alford, *British Economic Performance 1945–1975*, London, 1988, chapters 6 and 7.

25 Thomas K. McCraw, 'From partners to competitors', in McCraw, *America Versus Japan*, p. 13.

26 Keith Middlemas, *Power, Competition and the State*, Volume 2, London, 1990.

27 B. Elbaum and W. Lazonick (eds), *The Decline of the British Economy*, Oxford, 1986.

28 Elbaum and Lazonick, *Decline*, p. 11.

29 S. Tolliday, *Business, Banking and Politics: The Case of British Steel 1918–1939*, Cambridge, Mass., 1987. For similar examples from the cotton and coal industries, see J. H. Bamberg, 'The rationalisation of the British cotton industry in the interwar years', *Textile History*, XIX, 1988, pp. 83–101, and M. W. Kirby, *The British Coalmining Industry, 1870–1946: A Political and Economic History*, London, 1977.

30 D. C. Coleman, 'Failings and achievements: some British businesses 1910–80', *Business History*, XXIX, 1987, pp. 1–17.

31 Leslie Hannah, 'Strategy and structure in the manufacturing sector', in Leslie Hannah (ed.), *Management Strategy and Business Development*, London, 1976, pp. 184–202.

32 Michael Dintenfass, 'Entrepreneurial failure reconsidered: the case of the interwar British coal industry', *Business History Review*, LXII, 1988, pp. 1–34.

33 See, for example, E. Rothbarth, 'Causes of the superior efficiency of USA industry compared with British industry', *Economic Journal*, LVI, 1946; S. B. Saul (ed.), *Technological Change: The United States and Britain in the 19th Century*, London, 1970; Sue Bowden, 'Supply and demand constraints in the British car

industry in the interwar period: did the manufacturers get it right', University of Leeds Discussion Paper, 1989.

34 W. J. Reader, 'Imperial Chemical Industries and the state. 1926–1945', and L. Hannah, 'A pioneer of public enterprise: the Central Electricity Board and the National Grid, 1927–1940', in B. Supple (ed.), *Essays in British Business History*, Oxford, 1977; L. Hannah, *Electricity before Nationalisation*, London, 1979, especially chapter 4; T. C. Barker and Michael Robbins, *A History of London Transport*, Volume 2, London, 1974, chapters 15 and 16.

35 A. Slaven, 'Self-liquidation: The National Shipbuilders Security Ltd and British shipbuilding in the 1930s', in S. Palmer and G. Williams (eds) *Chartered and Unchartered Waters*, London, 1981.

36 Daniel C. Kramer, *State Capital and Private Enterprise: The Case of the UK National Enterprise Board*, London, 1989.

37 Robert R. Locke, *Management and Higher Education since 1940: The Influence of American and Japan on West Germany, Great Britain and France*, Cambridge, 1989; Michael Sanderson, 'Education and economic decline, 1890–1980s', *Oxford Review of Economic Policy* IV, 1988, pp. 38–50.

38 A. S. J. Baster, *The International Banks*, London, 1935, pp. 193–9; R. P. T. Davenport-Hines, *Dudley Docker*, Cambridge, 1984, pp. 137–49; W. A. Thomas, *The Finance of British Industry*, London, 1978, pp. 121–8; John Kinross, *Fifty Years in the City*, London, 1982, especially chapters 11 and 12.

39 Mancur Olson, *The Rise and Decline of Nations: Economic Growth, Stagnation and Social Rigidities*, New Haven and London, 1982.

40 Stephen Broadberry, 'The impact of the world wars on the long run performance of the British economy', *Oxford Review of Economic Policy*, IV, 1988, pp. 33–5.

41 Elbaum and Lazonick, *Decline*, p. 15.

42 Porter, *Competitive Advantage*, pp. 504–6.

43 Geoffrey Jones, 'The British government and foreign multinationals before 1970', in Martin Chick (ed.), *Government, Industries and Markets*, Aldershot, 1990.

44 Pollard, *Wasting*.

45 Locke, *Management and Higher Education since 1940*, pp. 227–30.

46 Peter A. Hall, *Governing the Economy. The Politics of State Intervention in Britain and France*, Cambridge, 1986; Geoffrey Ingham, *Capitalism Divided?*, London, 1984. See also Scott Newton and Dilwyn Porter, *Modernisation Frustrated*, London, 1988.

Productivity and competitive failure: British government policy and industry, 1914–19

Prior to 1914 government concern for the prosperity of British industry and the broader issue of national efficiency was intermittent. Alarmed by the potential political consequences of the so-called Great Depression of the 1880s and 1890s, the humiliation of the Boer War and growing Anglo-German rivalry, successive governments and commentators did show some awareness of the shortcomings of British industry.[1] Taking the period as a whole, however, state intervention amounted to little beyond the creation of an institutional environment conducive to the free operation of the market. Governments, both Liberal and Conservative, apart from interesting themselves in working conditions and social reform, showing passing interest in science and education and in introducing the Patent Act, had little coherent plan for industry.[2] Instead of a strategy for change governments evolved a series of tactics to deal with specific problems.

In imposing unprecedented logistic demands on the Bitish economy, the First World War gave rise to heightened awareness of productive deficiencies in a wide range of industries. On the eve of the war mass production techniques and the economies of large scale production, essential to service the material needs of modern warfare, were notable for their absence. Some industries vital for the prosecution of the war effort were either non-existent or grossly underdeveloped compared to their German counterparts. Many sectors of industry, moreover, were subject to trade union job controls which served to limit productive potential. In the circumstances of total war the state could not avoid assuming reponsibility for industrial modernisation. The creation of the Ministry of Munitions in 1915 was the most spectacular manifestation of government intervention to increase output in the armaments sector. Equally important was the Treasury Agreement of 1915, whereby the trade unions in the munitions-related industries agreed to lay down restrictive practices for the duration of the war. The results were spectacular in terms of productive achievement and there is certainly a case for claiming that the war years represented a major discontinuity in Britain's relative industrial decline. What is significant, however, in the context of the present paper, is that the experience of the war led to extensive analysis of Britain's economic prospects in peacetime. The Ministry of Reconstruction, founded in 1917, acted as the forcing ground for post-war planning, to the extent that the reconstruction of industry became a major objective of public policy. Yet despite the advances in industrial techniques achieved in consequence of state intervention, and a full appreciation of the competitive power of German and American enterprise, state-sponsored modernisation was stillborn. In explaining the failure of industrial reconstruction, historians have traditionally cited the post-war resurgence of political and economic orthodoxy as evidence of an unshaken faith in the pre-1914 liberal order. This is certainly part of the explanation for the demise of reconstruction, but it is only part of the story. The purpose of this paper, therefore, is to investigate more fully the circmstances which led to the abandonment of reconstructionist hopes. In the first section the case for reconstruction as it was expounded during the war is set out as a prelude to a discussion, in the second section, of the manifold constraints on state intervention to enhance industrial competitiveness. In the final section the paper analyses the circumstances which led to the full restoration of pre-war union job controls as a major constraint on productive efficiency.

1 The effects of the First World War on British industry

It was during the First World War that the problem of guaranteeing supplies of armaments, at a time when the supply of labour was dwindling, focused government attention on industrial efficiency. Forced by the logistic demands of the war, from 1916, to undertake unprecedented levels of state intervention, the wartime coalition can increasingly be said to have had an industrial policy worthy of the name. This was initially little more than a quest for maximum armament production yet, by the middle of 1917, it contained within it a vision for the future. The Ministry of Reconstruction, more often seen as a vehicle for social improvement and the source of a 'land fit for heroes' was, at its conception, intended to address the issue of industrial efficiency, especially where it related to competition with an undefeated Germany.[3] Its deliberations covered not only such matters as the structure of firms and the mechanisation of production, but also the need for co-operation between employers and employed, for extended industrial welfare and for improved scientific and technical education,[4] all cornerstones of efficient industrial production. The intention, therefore, was not to recreate the economic world of 1914 for, 'people [had] begun to doubt whether, after all, the social and industrial system of the country in pre-war days was so harmonious and so well organised as to deserve to be revived in its entirety'.[5]

Although war planning pre-dated 1914, attention had been directed towards military, especially naval, strategy rather than towards industry and its capacity to sustain supplies of armaments. Given the experience of the Boer War, this might seem strange and could be interpreted as a reflection of continued faith in free market forces by a Liberal government.[6] It has, however, been convincingly argued that both the neglect of industrial planning before 1914 and the 'business as usual' approach in the early part of the war were less the result of a slavish adherence to *laissez-faire* ideology than of the contemporary military strategy.[7] Before 1914 it was envisaged that a large-scale continental war would be both politically unpopular and logistically difficult to sustain. Instead it was intended that the British Expeditionary Force, operating in France, would be small and that the main instrument of war would be a naval blockade to paralyse the German economy. Existing armaments stocks, which had reached six months' supply by 1910, were deemed to be adequate for such a campaign, rendering detailed industrial planning superfluous. 'Business as usual', a phrase coined to reassure businessmen, was therefore considered adequate for a short campaign and to meet the immediate crisis. It was Lord Kitchener's plan for a vast army to fight a major land war, which rendered 'business as usual' redundant.[8] Nevertheless, that the Liberal government struggled on with a policy unequal to its task, until the 'shell scandal' of May 1915, did reflect some ideological divisions. At the same time it illustrates the difficulties of constructing a coherent policy when military and economic decision-making were separated.[9]

As the demand for munitions escalated, very real deficiencies in British industry, already apparent before 1914, were fully exposed. For example, British business was weak in such areas as chemicals, electrical engineering and modern metal manufacture.[10] Moreover, in certain industries crucial to the war effort, Germany had a virtual monopoly before 1914. Materials such as tungsten for steel, dyestuffs, optical glass and fine chemicals, all previously imported from Germany,[11] were soon in short supply. Indeed, a statement from the British Ignition Apparatus Association in 1918 admitted that:

> The British magneto industry has been entirely created since the war. Previous to that date practically the whole of supplies came from Germany and the large majority from the Bosch Company of Stuttgart who had established an almost complete monopoly. At the outbreak of war there was only one small firm making magnetos and their output was negligible.[12]

Supplies of products vital to the war effort were not the only problem. Early in the war, as the level of military recruitment grew, labour shortages emerged, especially of skilled labour and it soon became clear that it was not enough to prevent skilled workers from being swallowed up in the armed forces. The productivity of the existing labour force had to be raised and new sources of labour attracted.[13]

Government efforts to solve these deficiencies gathered momentum after the establishment of the Ministry of Munitions in June 1915.[14] Nevertheless, comparatively early in the war attempts had been made to counteract the worst effects of shortages of materials and manpower. Between August 1914 and March 1915 a series of three Acts for the *Defence of the Realm*[15] successively increased government control over industry. Under the third Act the government gained full power to take over any engineering factory and to control its processes.[16]

It was, however, in the sphere of labour supply and allocation that most progress was made before the middle of 1915. The problem of falling labour supply was a complex one. Not only was there a general shortage of labour at a time when demand was growing rapidly, but more critically there was a shortfall of skilled labour. It was an attempt to provide a solution to this difficulty that government policy began to veer towards greater specialisation and mechanisation of production. In this way, not only could productivity rise, but unconventional sources of labour could be used. The dilution of labour, attractive to both government and employers, was dependent upon a relaxation of trade union restrictive practices. These established rules and customs included a limitation on apprentices, restriction of output, regulation of overtime and curbs on recruitment. Long a thorn in the side of employers wishing to improve business performance, the waiving of these practices was seen as crucial to any substantial improvement in productivity.[17]

Negotiations began as early as August 1914 and continued until October 1915 with the introduction of the Dilution Scheme. In little over a year, trade unions in war-related industries had agreed that for the duration of the war, women could operate automatic machinery,[18] semi-skilled workers could carry out operations normally undertaken by skilled mechanics (Treasury Agreement 1915) and that the work of skilled men could be subdivided and simplified.[19] At the same time there were to be no strikes on war work and unions agreed to arbitration in any disputes.

The agreements were the foundation-stone upon which the achievement of higher productivity and efficiency rested. As such they represented not only a bargain between unions and government, but additionally with employers. Whilst it would be misguided to view them in terms of genuine consensus beyond the immediate requirements of the war effort, they nevertheless represented a significant achievement, however temporary and fragile, in the development of industrial relations. Trade union fears were lessened by government promises to 'undertake to use its influence to secure the restoration of previous conditions in every case after the war'[20] and also by employers undertaking to forgo profits resulting from the relaxation of union practices. All was given the force of law by the *Munitions of War Act*.[21] This legislation combined with the relaxation of the Factory Acts to allow for longer hours and for the labour of women and children, provided a firm basis for solving the labour supply problem.

If the relaxation of pre-war practices created the environment for change in war-related industries, it was the Ministry of Munitions which orchestrated the revolution in production methods. By the end of the war, the ministry had brought about significant changes in technology and business organisation. After 1916 electrical power was used in many factories for the first time, whilst encouragement was given to the use of the latest assembly line techniques in war industries. Management innovation, especially where it related to cost accounting and labour management was introduced.[22] To its credit the Ministry recognised that productivity depended on more than mechanisation. The potentially adverse effects of working excessively long hours were fully recognised, and welfare provision was actively promoted following the appointment of a Health and Munitions Workers' Committee to investigate the causes of industrial fatigue.[23]

The central importance of welfare in the ministry's labour strategy is illustrated by the establishment, in July 1916, of a permanent welfare section with B. S. Rowntree as its first director. He argued convincingly that:

> Real betterment of conditions springs in the last analysis from the conviction in the mind of the employer that here lies his plain duty, a duty which does not conflict with his business interests but promotes them, since it is obvious that workers who are in good health, and provided with the amenities of life, are more efficient workers.[24]

Under Rowntree's supervision, welfare officers were appointed and rest rooms, canteens and recreation facilities introduced. Although it is true that systematic welfare provision had been adopted in some large firms before 1914, the efforts of the Ministry of Munitions represented a significant advance in labour management.[25] It was therefore through state-sponsored technological and managerial innovation, that the wartime problems of production and supply were largely resolved.

Concern during the First World War was not only directed towards Britain's war-related industries, but also towards industrial performance generally and the adjustment of the economy to peace. In December 1915 Asquith appointed a Cabinet Reconstruction Committee to examine the likely capabilities of the British economy. Prompted by the fear that a strong Germany would conduct an effective trade war, in which Allied recovery efforts would fall victim to the dumping policies of German cartels and big business,[26] reconstruction involving continued state control was intended to sustain industrial efficiency in the event of a stalemate peace. As has been shown, the threat of German economic power had long been recognised outside government, but the war, by highlighting her industrial capabilities and the comparative weakness of British industry made it, initially, a central consideration for those involved in planning reconstruction.[27] That the war cabinet viewed the possible post-war threat from Germany with extreme apprehension is confirmed by the proposals made at the Paris Economic Conference of 1916. There the Entente powers agreed to launch an export drive against Germany, whilst at the same time pooling resources and aiming for economic independence.[28]

The initial inspiration for reconstruction may have been fear of Germany, but the aims and objectives of the Ministry of Reconstruction, established in August 1917, were more far-reaching. Its function, narrowly defined, was to: 'Consider and advise upon the problems which may arise out of the present war and may have to be dealt with on its termination'.[29] In a broad sense it was concerned with rebuilding 'the national life on a better and more enduring foundation',[30] or to use the phrase associated with Lloyd George, the creation of a 'land fit for heroes'.

One of the new ministry's principal objectives in this context was a desire to perpetuate the improvements in business organisation and industrial efficiency achieved in the munitions industries, into the post-war world. In this way, not only would the threat of future German competition be met, but through improved labour relations and working conditions social aims would be achieved.

Co-operation between employers and employees was seen as the single most important factor leading to improved industrial efficiency. It was argued that:

> The war had greatly altered the attitude of the worker ... that the returning soldier does not intend to acquiesce in the low paid conditions which characterized a part of British industry before the war, and that large changes in the relations of employers and employed are inevitable ... Increased output and improved conditions are two sides of the same shield. It is idle to hope to increase output unless the confidence of workers can be gained and their co-operation enlisted. Confidence must take the place of suspicion and public service the place of sectional self interest in the relations between parties.[31]

In a similar vein the *Committee on Commercial and Industrial Policy after the War* concluded:

> It is in our opinion a matter of vital importance that, alike in the old established industries and the new branches of manufacture which have arisen during the war, both employers and employed should make every effort to attain the largest possible volume of production by increased efficiency of industrial organization and processes, by more intensive working and by the adoption of the best and most economical methods of distribution.[32]

It was in order to find a vehicle for such co-operation that the Whitley Committee was set up in 1916. Taking its name from its chairman John Whitley, the remit of the *Committee on Relations between Employers and Employed* was:

> To make and consider suggestions for securing a permanent improvement between employers and workmen. [and] To recommend means for securing that industrial

conditions affecting the relations between employers and workmen shall be systematically reviewed by those concerned, with a view to improving conditions in the future.[33]

The Committee concluded that these objectives could be best achieved if there were organisations which allowed for free discussion between trade unions and employers.

In order to avoid all types of disputes a three-tier system of Joint Industrial Councils was proposed, composed of both union and employers' representatives. The National Joint Industrial Council for an industry was to consider the interests of the industry as a whole. At a local level there were to be district councils with a sub-structure of works committees to settle minor difficulties and 'promote organization within the individual workshop, mine or factory.'[34] The briefs of these councils were to be wide-ranging and included securing co-operation on such issues as wages and working conditions, regulation of production, perfecting the products of the industry, whilst at the same time providing a mechanism for settling differences.[35] Thus the Joint Industrial Councils were the pivot upon which plans for improving industrial efficiency rested. At the same time they offered a potential solution to the problem of the growing centralization of the unions.

Reaction to the proposed Joint Industrial Councils was mixed. Some employers had long favoured such a development, with proposals dating back to 1910.[36] Union reaction, on the other hand, was uncertain ranging from suspicion from some of the skilled unions, to enthusiasm from unskilled unions and those in industries where there was little formal collective bargaining.[37] By 1918 twenty-six Joint Industrial Councils were in existence in industries as diverse as motor cars, chemicals, bobbins and pottery, with fourteen more at the stage of draft constitutions.[38]

It is important to note, however, that in the six larger staple trades, including shipbuilding, cotton, coalmining and railways, little progress had been made. It was argued that:

> In each of these cases special difficulties exist, and it has been found, not unnaturally, the larger the industry, the greater the difficulties that have to be overcome, owing to the complexity of its existing organization and the difficulties of reconciling the views of the various bodies concerned, where they have not previously been accustomed to working together. It is therefore to be expected that the largest industries will be the last to adapt themselves to the new scheme.[39]

As the following section will show, the 'special difficulties' included a level of antagonism between labour and capital which was entirely inconsistent with the philosophy of Whitleyism. Nevertheless, before the end of the war, some independent progress had been made, in the shape of the Cotton Control Board, for example.

Thus plans for reconstruction reflected a profound desire for the resolution of conflict between labour and capital. At the same time it was assumed that in the post-war world there would be far greater co-operation between firms. It was, for example, proposed that within industries there should be co-operation and consolidation in securing supplies, in standardising production, in scientific and industrial research, and in marketing.[40] Concern was also expressed at the potential for monopoly pricing which might follow the end of the war, in view of the increase in the number of trade associations in British industry after 1914. Thus a standing Committee on Trusts was established under the auspices of the Ministry of Reconstruction, whose report in 1919 was to lead to the Profiteering Act of that year.[41]

The hallmark of the government's proposals for economic reconstruction was that they were an attempt to build upon a new level of consultation between government, employers and unions. Lloyd George's tactic of involving leading industrialists such as Eric Geddes to participate in framing labour and production policies, was a shrewd one.[42] Initially, at least, and with no end to the war in sight, it created an image of consensus. At the same time the rapid expansion of the Federation of British Industries (FBI) to reach 793 firms by 1918, gave the government a useful counterbalance to the unions. As the next section will reveal, however, it would be wrong to assume that the new found consensus between government, employers and unions was of sufficient depth to be sustained beyond the end of the war. Indeed, even during the war, especially in the later stages, labour unrest was not uncommon.[43] Moreover, whilst government plans for industry may have reflected the ideas of industrialists, they were derived, it seems, from a vocal minority of productioneers and trade warriors who were not representative of the majority of businessmen.

One of the most strident and influential businessmen during the war years was Dudley Docker. Resenting the decline of the British industrial economy, he had argued, even before the war, that 'we must produce as we have never produced before and larger production means larger wages'.[44] A leading productioneer after 1914, he viewed the war as British industry's last chance to overhaul its German and American rivals.[45] He and the other productioneers advanced prescriptions for revitalising British industry which clearly influenced those in government concerned with reconstruction. Indeed, in May 1917, Docker strenuously argued for the creation of a Ministry of Reconstruction which would be a vehicle for the establishment of the corporatist business world he favoured.[46] Always fearful of the potential threat from Germany, he argued not only for an adequate forum for employers and employed, but also for the rationalisation and modernisation of industry, supported by adequate scientific research and protected by an Imperial Tariff.[47] As founder of the FBI Docker was in a strong position to influence government and he played a key role in helping to launch the Joint Industrial Councils in 1917.[48] He and his fellow productioneers dominated the FBI in its early years, but as the following section will point out, productioneers were hardly representative of businessmen as a whole.

It is fair to say that the First World War marked a major step forward in government thinking towards industry. The achievement of maximum productivity as opposed to simply increasing output, became a major policy objective. This concern with industrial efficiency, which increasingly embraced an awareness of the importance of welfare and labour conditions, was also accompanied by greater co-operation between governments, employers and employed. It is, however, worth observing that, as in the pre-war period, part at least of the focus of government policy came from the immediate problems of the war and its likely aftermath. As the following section will demonstrate, the solution or removal of these problems reduced much of the momentum for change and, with it, the commitment of the government to a radical programme of reconstruction.

2 The failure of reconstruction after the war

In his study of the abandonment of wartime controls after 1918, Tawney lamented the lack of any intellectual conversion to the merits of a greater degree of government participation in peacetime economic affairs.[49] Writing in 1943, in the midst of a war in which state intervention was far more extensive and rigorously applied, it was understandable, given his collectivist ideals, that he should wish to ensure that post-war reconstruction was in no way inhibited by a return to some mystical conception of 'normalcy'. Yet as an objective observer, Tawney was willing to concede that the rapid dismantling of controls after 1918, however much it was to be regretted, was none the less inevitable. The failure of reconstruction was a direct consequence of the conspicuous failure of government 'to inform itself and the public as to the merits and defects of different war controls, or to discriminate between those whose utility had ceased and those which could with advantage be retained'.[50] Although it has been argued more recently that, but for a reprehensible degree of financial parsimony on the part of the Treasury, reformist influences were as strong as those in favour of reaction in the immediate post-war years,[51] the image presented by Tawney of a rudderless administration, anxious to divest itself of the apparatus of intervention in the face of a resurgence of orthodox economic opinion within the business and financial community, remains substantially intact.

Whilst there is substance in Tawney's argument, as an all-embracing explanation of the failure of reconstruction it is excessively monocausal. Hesitancies and contradictions in government policy on reconstruction were certainly present at the end of the war, especially in the context of demobilisation, but these must be set alongside the evolving situation in labour relations, divisions of opinion and strategy within the business community, the impact on government policy of an unexpectedly rapid termination of the war, and the restoration of Treasury control over the Whitehall administrative machine.

As noted in the previous section, the view from within the Ministry of Reconstruction was that the process of reconstruction itself was mainly dependent upon a successful *rapprochement* between capital and organised labour — hence the emphasis on the formation of joint industrial councils focusing on the Whitley system. Yet as the Ministry of Labour recognised at an early stage, Whitleyism was 'fatally flawed'.[52] Within the trade union movement the craft unions feared that Whitleyism would lead to combinations between semi-skilled workers and employers, to the detriment of their privileged position in the hierarchy of industrial relations. A further conflict of interest centred on the disputed authority between shop stewards and national trade union organisations. To the extent that centrally-appointed district committees provided the focus for local control, what role would be left for the projected works committees? There was also a potential conflict of purpose between unions and employers as to the use of Whitley Councils. As a result of their wartime experience, the unions had embraced the concept of workers' control as a means of sharing in industrial management. Sympathetic and progressive employers, however, had proceeded only to the point of accepting the principle of mutual co-operation to raise productivity, and it was this version of industrial partnership that was embodied in the original Whitley Report. Whitleyism, therefore, fell short of meeting the heightened aspirations of wartime labour.

Above all, Whitleyism was regarded with deep suspicion and scepticism by key sections of employers and workers.[53] In the former case the centre of opposition lay in the hardline Engineering Employers' Federation (EEF). Its secretary was Sir Allan Smith, described by Walter Citrine as 'an expert in procrastination, whose icy cold speeches were enough to freeze to death any progress his members might feel'.[54] Referring habitually to trade unions as 'the enemy', he was hardly an enthusiastic supporter of industrial consensus.[55] Complementing the engineering employers was that 'weapon of unpredictable power' the Triple Industrial Alliance, composed of the Miners' Federation of Great Britain (MFGB), the National Union of Railwaymen (NUR) and the Transport Workers' Federation (TWF). In 1919 the NUR's *Railway Review* stated that 'no useful purpose is served by collusion with the employer through the Government to maintain the existing order of society' (27 June 1919). Such sentiments struck at the very roots of reconstruction, but the most uncompromising rejection of industrial consensus was to be found within the coal industry. Even before 1914, two mutually incompatible views of the industry had emerged, with the miners favouring national wage settlements whilst colliery owners remained committed to the industry's decentralised collective bargaining structure. The national pooling of profits during the First World War drove the two sides even further apart.[56] As William Brace, the Welsh miners' leader stated in October 1919, 'The war has driven us at least twenty-five years in advance of where we were in thought in 1914. The young men have thought deeply and indeed, they are educated.'[57] It was the belief that state control represented a dramatic and irreversible advance in their position which helps to explain the apparent audacity of the miners' post-war programme, published as a series of demands in January 1919. In addition to a 30 per cent increase in gross wages (excluding wartime increases) and a six-hour day, it included a demand for the nationalisation of mines which contained the novel stipulation that there should be 'joint control by the workmen and the state' — clearly a reflection of the miners' experience of wartime control. In combining political with economic demands, the MFGB had moved far beyond any notion of industrial consensus based on a partnership between labour and capital.

The *coup de grâce* for reconstructionist hopes based upon the principle of consensus was administered by the National Industrial Conference of trade union and employers' representatives which met under government auspices in February 1919. Summoned by Lloyd George to give form and content to the proposals for the limitation of the working week and minimum rates of pay the detailed schemes drawn up by the Provisional Joint Committee (PJC) were stillborn.[58] Although many of the PJC's minor recommendations were implemented, the government baulked at surrendering executive responsibility to a proposed 600-strong national industrial council or parliament. As Sir Robert Horne, the Minister of Labour, stated in September 1919, it was 'impossible for the government to surrender its freedom to any body of people however eminent'.[59] It was the constitutional issue which provided the key to the failure of the conference, for in the final analysis:

> neither employers nor union leaders were prepared to defend and fight for consensus policies — and thereby risk their standing with the rank and file — unless they were guaranteed in return some concession of real power from the government.[60]

It has also been pointed out that contrary to contemporary belief, the conference did not represent 'the consummation of the Whitley movement'. In 1919 Whitleyism was making some headway in a wide range of industries with negotiations for joint councils under way, even in engineering and coalmining. There was, of course, the threat that trade boards would be imposed upon weakly organised industries, complemented by the promise that they would become the established representative body for an industry. As Rodney Lowe has emphasised, however, the representatives of Whitleyism in attendance at the conference were all but ignored, as was the movement's potential role in the proposed industrial parliament. In this sense the conference was 'a political betrayal of Whitleyism' of such magnitude that its growth was permanently stunted. Negotiations in leading trades were soon terminated and by the early 1920s government too had lost interest.[61]

Among employers it has been noted that the group most in favour of centralised reconstruction plans were the productioneers, who played the key role in the formation of the FBI as a 'Business Parliament' to co-operate with government and finance in formulating rationalisation and export strategies.[62] Yet even at the peak of their influence in 1917–18, the productioneers were in open conflict with other powerful business groups within the industrial community. Their espousal of tariff protection as the principal weapon in combating German industrial power was hardly likely to appeal to *laissez-faire* interests in Lancashire and elsewhere, whilst the advocacy of works committees provoked a direct clash with shipbuilding and engineering employers. Indeed, the EEF engaged in a running battle with the Labour Committee of the FBI which eventually resulted in the FBI being deprived of any influence in the area of industrial relations, a development which was consolidated by the formation, in 1919 of the National Confederation of Employers' Organisations under the auspices of Sir Allan Smith.[63] The founding of the Imperial Association of Commerce in July 1918, under the presidency of Lord Inchcape, was a further blow both to the productioneers within the FBI and to the reconstruction movement in general. Disturbed at the activities of such arch-trade warriors as Dudley Docker, Inchcape and his associates in shipping and financial circles were determined to undermine the campaign for corporatist reform by proclaiming the virtues of the *status quo* in British business and commerce. Protection would 'crush and destroy that splendid initiative and individual enterprise which has in the past so largely contributed to the strength of the Empire'. As Davenport-Hines has pointed out, the Association's rhetoric was final 'proof ... that unanimity was impossible among the mass of British businessmen and its formation foreshadowed the destruction by the early 1920s of all ... hopes for corporatist reform'.[64]

The failure of consensus between capital and labour and within the business sector was thus a powerful factor leading to the demise of reconstruction in all its aspects, both economic and social. There was, however, an even more fundamental cause which was especially damaging to those who favoured reconstruction as a means to bolster industrial efficiency. As Cline has argued, to discuss the abandonment of reconstructionist hopes without reference to the issue of national security — an omission perpetrated by virtually all contributors to the relevant literature since Tawney — is to ignore the fundamental point that the process of decontrol was precipitated by Germany's military collapse.[65] As Cline has stated:

> Germany's sudden collapse in the closing months of the war removed the strongest, least vulnerable justification for the expanded state; it removed the prop which would have sustained the programme of state-initiated economic development.[66]

On this interpretation, reconstruction was foiled for the simple reason that it was perceived to be unnecessary. The main pre-war threat to Britain's economic hegemony had been defeated and whilst physical destruction of the German economy was minimal, the Allied powers proceeded, during the course of 1919, to impose a 'Carthaginian' peace settlement, which appeared to wreak havoc with the German economy. The loss of key raw materials in Alsace-Lorraine, the surrender of colonies and of a considerable portion of the merchant marine could be viewed as crippling blows to the material and commercial base of German industry — such was the view of Keynes — and this takes no account of the reparations settlements imposed on the defeated powers to the detriment of their external and internal financial stability.[67]

There was another major sense in which the end of the war paved the way for the demise of reconstruction. Concern at the growth of public expenditure had been increasingly evident in 1917 and 1918, when the parliamentary Select Committee on National Expenditure highlighted the lack of Treasury control both over the staffing levels of new ministries, such as Labour, Food and Shipping and the domestic spending of the War Office and the Ministry of Munitions.[68] After the Armistice the chorus of complaint about excessive public expenditure began to mount, reaching a crescendo during the summer of 1919, coinciding with the debates on the Finance Bill in the House of Commons.[69] The impression was created that whilst the Chancellor of the Exchequer, Austen Chamberlain, was attempting to control public expenditure, he was not being helped by his colleagues in spending departments, not least in the Ministry of Reconstruction. In July 1919 Chamberlain warned his Cabinet colleagues that if expenditure was not reined back the country faced financial ruin, in the light of the inflationary potential of a vastly increased floating debt. This was not just a threat to budgetary stability, but also to the declared policy of returning to the gold standard as soon as practicably possible. In August Chamberlain was given authority to vet the financial implications of all parliamentary bills, a step which foreshadowed the reorganisation of the Treasury, with the creation of the Establishments Division to control civil service pay and organisation. At the same time the Permanent Secretary to the Treasury was designated head of the civil service and Permanent Secretary to the Prime Minister as First Lord of the Treasury.[70] The stage was thus set for the Geddes Axe and the final extinction of reconstructionist hopes.

3 Organised labour after 1918

The previous section has drawn attention to several factors which, cumulatively, paved the way for decontrol. There remains to be considered a final issue, also related to the end of the war, which was never satisfactorily resolved by reconstructionists and which had major implications for the productive efficiency of British industry in the long term. As indicated earlier, by signing the Treasury Agreement of March 1915, the trade unions had agreed to the abandonment of restrictive working practices, in return for a government commitment to control profiteering. The wording of the agreement appeared to be unequivocal in meaning, as the following extract shows:

Any departure during the war from the practice ruling in our workshops, shipyards and other industries prior to the war, shall only be for the period of the war.

No change in practice made during the war shall be allowed to prejudice the position of the workpeople in our employment, or of their trade unions in regard to the resumption and maintenance after the war of any rules or customs prior to the war.

In any readjustment of staff which may have to be effected after the war, priority of employment will be given to workmen in our employment at the beginning of the war who are serving with the colours or who are now in employment.[71]

The relaxation of trade practices was thus for the duration of the war only, and it is an indication of ministerial awareness of trade union suspicions in this regard that members of the War Cabinet and other government spokesmen felt bound to offer repeated assurances that the government 'had pledged themselves up to the hilt to restore the pre-war conditions'.[72] In the summer of 1918 Sir Stephenson Kent, Director General of Demobilisation, was appointed to chair an interdepartmental conference on the redemption of the pledges to labour, and in a report addressed to the Ministry of Reconstruction he concluded that the preparation of a draft bill was essential as a means to placate mounting labour unrest.

The prospect of an early and unconditional redemption of the pledges enshrined in the Treasury Agreement sparked off a furious debate in Whitehall. Winston Churchill, as Minister of Munitions, deplored the speedy restoration of pre-war practices on the grounds that it could vastly complicate the business of demobilisation. Dilutees in the munitions-related industries would be the chief casualties and as their jobs were terminated the level of unemployment would rise substantially.[73] It was essential, therefore, that the government should 'retain control of industrial conditions' during the transition from war to peace. To restore union privileges would also 'entrench a number of small and close corporations in restraint of trade'.[74] It was this latter point, raising the issue of industrial efficiency, which was highlighted by the President of the Board of Trade, Sir Albert Stanley. A businessman drafted into government by Lloyd George, as part of his 'experiment',[75] Stanley's views were reminiscent of productioneering sentiment. A government bill to give effect to the redemption of pledges would be interpreted as official connivance at the restriction of output and would inevitably be 'contrasted with the repeated declarations of Ministers that our only hope of economic salvation after the war is to secure the maximum of output'.[76] The most blistering attack on restoration, however, was launched by Lord Weir, another Lloyd George appointee from the world of business. In a powerfully worded note Weir countered the argument that the government was honour-bound to redeem its pledges on the grounds that wartime labour had not fulfilled its side of the bargain. There was even at this late stage in the war, 'active hostility to dilution, refusals to work on any system of payment by results, [and] refusals to withdraw demarcation restrictions.' The right to strike had also not been abandoned. The heart of Weir's attack, however, was that restoration of pre-war practices would seriously damage the future prospects of British industry:

If there is one thing clear and definite in regard to pre-war conditions it is surely the necessity of leaving industry untrammelled by precedents or practices which detrimentally affect efficiency of production, and in this respect, the present Bill if passed, would represent the industrial suicide of Britain. There is surely evidence that British industry will be faced with the heaviest foreign and allied competition, that the supply of capital will be lacking, and that industrial risks will be so great as to restrain development, unless a stable and efficient set of conditions is created.[77]

After drawing a link between higher productivity and high wages, Weir then proceeded to highlight the growing competitive power of American industry, before appealing for detailed investigation into the effects of restrictive practices in the engineering industry.

The case for reneging on the pledges was thus argued powerfully and without ambiguity. In the light of Weir's warnings, the fact that the government proceeded with legislation can be regarded as one of the greatest lost opportunities to free industry of the deadweight of 'ca canny' and union job controls, which served to reduce productivity. This is, however, a hindsight argument, and it is important to note that an equally vigorous and ultimately more persuasive case was presented by the supporters of restoration. It is hardly surprising that the most vocal support for restoration was to be found within the Ministry of Reconstruction. The minister, Christopher Addison, was a convinced believer in Whitleyism and the creation of joint councils generally in industry. Addison argued that a breach of the original Treasury Agreement would, indeed, involve 'the honour of the government', but the moral obligation to the trade unions was buttressed by further arguments which were to prove decisive. Of greatest importance was the fact that restoration was inextricably bound up with reconstruction. As Addison pointed out:

almost every avenue towards industrial reconstruction is at present a *cul de sac*, blocked by the restoration question. At every turn in my efforts to deal with the many complicated labour and industrial problems arising out of the war, I am brought face to face with an unfulfilled promise and a series of questions which cannot be settled until the promise is redeemed. The future of women in industry, the place of the semi-skilled worker, the development of improved methods, increased productivity, all depend in the last resort upon the attitudes of the trade unions of skilled workers, which will be determined by the action of the government with regard to the restoration of pre-war practices.[78]

In combating the views of Stanley and Weir, Addison was adamant that refusal to honour the pledge would create havoc in labour relations, thereby undermining the post-war reconstruction of industry. The 'stillborn opposition of the skilled unions would be such as to make the establishment of working rules more in harmony with the new conditions and opportunities [ushered in by war] impossible'. In any event, it was unlikely that trade unionists would insist on literal restoration, viewing it as a bargaining counter.[79] But undue hesitation on the part of government:

would be tantamount to throwing over the accredited leaders of the unions, who would be charged with having betrayed the men. The upshot would be the destruction of organised trade unionism, and a great stimulus to the extremists.[80]

Reinforcing Addison's views was a direct appeal to the War Cabinet from the Ministry of Labour in which Sir David Shackleton, the Permanent Secretary, pointed to the uncertainty concerning restoration as a potent factor in stimulating unrest. Employers in general supported restoration as the best guarantee of isolating the unofficial far left and extremist elements within the trade unions.[81]

After protracted negotiations between employers and unions a bill to restore pre-war practices was enacted in August 1919. Its provisions went further than many had expected, especially in the requirement, under threat of financial penalty, for employers in newly created establishments, or those which had taken up munitions production during the war, to permit the introduction of practices 'as obtained before the war' in similar job situations. Between 30,000 and 40,000 pre-war trade restrictions had been laid down under the Treasury Agreement and as G. D. H. Cole pointed out, the provisions of the Act were implemented 'rapidly and smoothly and with relatively little friction',[82] thus suggesting that employers preferred 'in the face of immediate demand, to avoid trouble, to revert to the old methods and to get back their former staffs, rather than engage in the hazardous enterprise' of radical workshop reorganisation to improve their competitive position in peacetime markets.[83]

To argue that the government, with or without the backing of employers, should have seized the opportunity provided by the war to bring about substantial reform in labour relations consistent with the attainment of high productivity and export competitiveness is to ignore contemporary political realities and the pressure of immediate circumstances. In the autumn and winter of 1918 organised labour was in the ascendant and to paraphrase Mowat it had yet to be 'foiled'. Labour unrest was mounting and this boded ill for mass demobilisation. Even Churchill was forced to admit in the midst of a strike by railway clerks in February 1919 that:

> Trade union organisation was very important, and the more moderate its officials were, the less representative it was; but it was the only organisation with which the government could deal. The curse of trade unionism was that there was not enough of it and it was not highly enough developed to make its branch secretaries fall into line with the head office. With a powerful trade union, either peace or war could be made.[84]

Above all, perhaps, reconstructionist hopes were still alive and the apparently retrograde step of returning to the pre-1914 *status quo* in shop-floor organisation could still be viewed in an optimistic light.

4 Conclusion

In conclusion, it seems clear that a broader view of the failure of industrial reconstruction is necessary. The movement for a 'return to normalcy' and a recommitment to liberal ideology at the end of the war were important factors, but they interacted in a complex way with other powerful influences to place impediments in the way of closer government–industry relations in peacetime. It is true that the policy of the Ministry of Munitions in encouraging mergers and the adoption of best practice techniques over wide sectors of the engineering industry can be regarded as an important progenitor of the rationalisation movement of the later 1920s. It is also the case that many of the businessmen who were to support the movement towards large-scale amalgamation in British industry first became alerted to the possibilities of scale economies as a result of their experience as administrators in wartime production departments. But these are hindsight observations. The fact remains that industrial concensus and a vigorous state-sponsored drive for business efficiency embracing scientific management, standardisation, and the adoption of high level technologies were entirely dependent upon expectations of a particular outcome of the war. With the military collapse of Germany, reconstruction in all of its wartime manifestations lost its essential foundation. The war economy may have been a remarkable achievement, but it was hardly likely to be viewed as a suitable model for the post-war world when its logistic and organisational successes were entirely dependent upon the abandonment of cost constraints at a time when the nation was fighting for survival.

Notes

1 *Final Report from the Royal Commission on the Depression in Trade and Industry with Minutes of Evidence and Appendices*, 1896, Cd. 4893; E. Williams, *Made in Germany*, London, 1896; W. H. Dawson, 'Germany's commercial progress', *Economic Journal*, XI, 1901, pp. 565–75; A. Shadwell *Industrial Efficiency: A Comparative Study of Industrial Life in England, Germany and America*, London, 1906; G. R. Searle, *The Quest for National Efficiency*, Berkeley

and Los Angeles, 1971, pp. 8–9; S. Newton and D. Porter, *Modernization Frustrated: The Politics of Industrial Decline in Britain Since 1900*, London, 1988, p. 12; B. Semmel, *Imperialism and Social Reform: English Imperial Thought 1895–1914*, p. 43.

2 C. J. Wrigley, 'The government and industrial relations', in C. J. Wrigley, *A History of British Industrial Relations, Volume I 1875–1914*, Brighton, 1982, pp. 135–58.

3 Peter Cline, 'Winding down the war economy: British plans for peacetime recovery, 1916–19', in Kathleen Burk (ed.), *War and the State: The Transformation of British Government 1914–19*, London, 1982, p. 159.

4 Ministry of Reconstruction, *Problems of Reconstruction*, Pamphlet I, *The Aims of Reconstruction*, London, 1918.

5 *Ibid*., p. 3.

6 S. J. Hurwitz, *State Intervention in Great Britain: a Study of Economic Control and Social Response 1914–19*, London, 1968, pp. 61–88.

7 D. French, 'Business as usual', in Burk, *War and the State*, pp. 7–13.

8 *Ibid*., p. 8.

9 *Ibid*., p. 26.

10 B. W. E. Alford, 'Lost opportunities: British business and businessmen during the First World War', in N. McKendrick and R. P. Outhwaite (eds), *Business Life and Public Policy: Essays in Honour of D. C. Coleman*, Cambridge, 1986, p. 206.

11 *Final Report of the Committee on Commercial and Industrial Policy after the War*, 1918, Cd. 9035, p. 94.

12 PRO, BT55, Engineering Trades Committee, p. 27.

13 Hurwitz, *State Intervention*, p. 89.

14 5 & 6 Geo5 c51, *An Act for Establishing in Connection with the Present War a Ministry of Munitions of War and for Purposes Incidental Thereto.*

15 4 & 5 Geo5 c29, *Defense of the Realm Act*; 5 Geo5 c8, *An Act to Consolidate and Amend the Defense of the Realm Act*; 5 Geo5 c23, *An Act to Amend the Defense of the Realm Consolidation Act.*

16 5 Geo5 c23, *An Act to Amend the Defense of the Realm Consolidation Act.*

17 Hurwitz, *State Intervention*, p. 89.

18 G. D. H. Cole, *Trade Unionism and Munitions*, Oxford, 1925, pp. 67–9.

19 Hurwitz,*State Intervention*, p. 97

20 Quoted in Cole, *Trade Unionism and Munitions*, p. 71.

21 5 & 6 Geo5 c51, *An Act for Establishing in Connection with the Present War a Ministry of Munitions of War and for Purposes Incidental Thereto.*

22 Wrigley, 'The government and industrial relations', pp. 47–9.

23 Hurwitz, *State Intervention*, p. 113.

24 Ministry of Munitions, *Official History*, London, 1920, Volume V, p. 12.

25 R. Fitzgerald, *British Labour Management and Industrial Welfare 1846–1939*, London, 1987, pp. 180–4.

26 PRO, MUN 4/6473.

27 Cline, 'Winding down the war economy', pp. 159–64.

28 V. H. Rothwell, *British War Aims and Peace Diplomacy 1914–18*, Oxford, 1971, p. 269.

29 Ministry of Reconstruction, *Problems of Reconstruction*, p. 5.

30 *Ibid*., p. 4.

31 *Ibid*., p. 10.

32 *Final Report of the Committee on Commercial and Industrial Policy after the War*, 1918, Cd. 9035, p. 22.

33 Ministry of Reconstruction, *Problems of Reconstruction*, Pamphlet 18, *Industrial Councils: the Whitley Scheme*, London, 1918, p. 1.

34 *Ibid*., p. 3.

35 *Ibid*., p. 3.

36 C. J. Wrigley, 'The First World War and state intervention in industrial relations', in C. J. Wrigley (ed.), *A History of British Industrial Relations, Volume II, 1914–39*, Brighton, 1987, p. 55.

37 *Ibid*., p. 59.

38 Ministry of Reconstruction, *Problems of Reconstruction*, Pamphlet 18, p. 8.

39 *Ibid*., p. 9.

40 *Final Report of the Committee on Commercial and Industrial Policy after the War*, 1918, Cd. 9035, p. 34.

41 Ministry of Reconstruction, *Report of the Committee on Trusts*, 1919, Cd. 9236.

42 C. J. Wrigley, 'The First World War', p. 51; K. Grieves, *Sir Eric Geddes: Business and Government in War and Peace*, Manchester, 1989.

43 Hurwitz, *State Intervention*, pp. 243–85.

44 R. P. T. Davenport-Hines, *Dudley Docker: The Life and Times of a Trade Warrior*, Cambridge, 1984, p. 2.

45 *Ibid*., p. 84.

46 *Ibid*., pp. 99–100.

47 *Ibid*., pp. 4, 99–100.

48 *Ibid*., p. 116.

49 R. H. Tawney, 'The abolition of economic controls, 1918–21', in J. M. Winter (ed.), *History and Society: Essays by R. H. Tawney*, London, 1978, pp. 129–86.

50 *Ibid*., p. 149.

51 Rodney Lowe, 'The erosion of state intervention in Britain, 1917–24', *Economic History Review*, XXXI, 1979, p. 271.

52 Rodney Lowe, *Adjusting to Democracy: The Role of the Ministry of Labour in British Politics, 1916–1939*, Oxford, 1986, p. 92.

53 Alan Fox, *History and Heritage: The Social Origins of the British Industrial Relations System*, London, 1985, p. 296.

54 Walter Citrine, *Men and Work*, London, 1964, p. 247.

55 Lowe, *Adjusting to Democracy*, pp. 83–4.

56 M. W. Kirby, *The British Coalmining Industry, 1870–194e: A Political and Economic History*, London, 1977, pp. 24–48.

57 PRO, CAB24/90, 9 October 1919.

58 Rodney Lowe, 'The failure of consensus in Britain: the National Industrial Conference, 1919–21', *Historical Journal*, XXI, 1978, pp. 649–75.

59 PRO, CAB24/92, CP25, n. d.

60 Lowe, 'The failure of consensus', p. 658.

61 *Ibid*., pp. 672–3.

62 John Turner, 'The politics of organised business in the First World War', in

John Turner (ed.), *Businessmen and Politics: Studies of Business Activity in Britain 1900–45*, London, 1984, pp. 33–49.

63 Davenport-Hines, *Dudley Docker*, pp. 114–16.

64 *Ibid.*, p. 119.

65 P. Cline, 'Winding down the war economy', pp. 157–81.

66 *Ibid.*, p. 159.

67 J. M. Keynes, *The Economic Consequences of the Peace*, London, 1919.

68 *First Report from the Select Committee on National Expenditure: Second Report from the Select Committee on National Expenditure*, Parliamentary Papers, 1917–18.

69 Hansard, *House of Commons Debates*, 5th series, vol. 116, cols 2133, 2147, 4 June 1919.

70 Kathleen Burk, 'The Treasury: from impotence to power', in Kathleen Burk (ed.), *War and the State*, pp. 84–107.

71 PRO, RECO 1/800, 'Acceleration of output on government work', 19 March 1915.

72 Hansard. *House of Lords Debates*, vol. 907, col. 1923, 7 November 1917.

73 PRO CAB 23/8, WC 487, 16 October 1918.

74 *Ibid.*

75 Peter Cline, 'Eric Geddes and the "experiment" with businessmen in government', in K. D. Brown (ed.), *Essays in Anti-Labour History: Responses to the Rise of Labour in Britain*, London, 1974, pp. 74–104.

76 PRO, CAB 24/63, GT 5957, 10 October 1918.

77 PRO, CAB 24/63, GT 5974, *Note by Lord Weir on the Restoration of Pre-War Practices Bill*, 11 October 1918.

78 PRO, RECO 1/800, *The Restoration of Pre-War Practices Bill*, 12 October 1918.

79 Gerry R. Rubin, 'Law as a bargaining weapon: British Labour and the Restoration of Pre-War Practices Act, 1919', *Historical Journal*, XXXII, 1989, p. 932.

80 PRO, CAB 23/8, WC 491, 19 October 1918.

81 P. B. Johnson, *Land Fit for Heroes*, Chicago, 1968, p. 266.

82 G. D. H. Cole, *Trade Unionism and Munitions*, Oxford, 1923, pp. 124–5.

83 Sidney and Beatrice Webb, *The History of Trade Unionism*, London, 1920, p. 643.

84 Johnson, op. cit., p. 266.

3 *Jim Tomlinson*

A missed opportunity? Labour and the productivity problem, 1945–51[1]

The problem of industrial competitiveness has many facets, both of analysis and possible solution. One sub-division of the issue would be threefold, into costs, productivity, and the exchange rate.[2] At different times different elements of this sub-division have had prominence in discussion and policy-making. This chapter focuses on the period of the Labour governments of the immediate post-1945 years, when it was productivity which was central to the government's economic policy agenda.

The priority accorded to productivity by the Attlee government does not perhaps match popular perceptions of that government, but it is clear in recent scholarly work.[3] In similar fashion to the effects of the First World War discussed by Maurice Kirby and Mary Rose in Chapter 2, the Second World War exposed the deficiencies of Britain's industry to powerful scrutiny and some reform.[4] For the Labour government, this same problem rapidly loomed large as it became evident that the major threat to its policies of full employment and economic stability came not from deflation, but from an incapacity of the economy to provide the exports to finance import levels and the other drains on foreign exchange.[5] As the economy was at full employment, and as demobilisation was completed, there was a simple logic in arguing that the only possible source of extra output was enhanced productivity.

The problem of Britain's balance of payments, it was generally accepted, should not be addressed by varying the exchange rate, at least not in the short run. There was no lack of demand for British exports in the immediate post-war period, but a lack of supply. Equally costs, especially wage costs, were not inflating rapidly in this period, partly because of the government's successful incomes policy in 1948–50.[6] Hence, of the trinity of components of competitiveness, productivity came to the fore.

Implicit in the approach of this paper is the view that government policy may aid productivity. The alternative view, that government attempts to raise productivity are bound to be self-defeating and indeed harmful, may be argued from first principles — from a belief that all government intervention in industry cuts across the efficiency-generating characteristics of the private sector. At this level of generality the argument — 'the market versus state intervention' — seems to be one more suited to the nursery than to academic discussion. Ultimately this dichotomy rests on a belief that markets and states are uniform in their effects, independent of the institutional, political and other contexts in which they exist. The focus here will be at a much more specific level, and will be concerned with the relation in this period between a particular government's welfare objectives — full employment and the creation of the welfare state — and the desire to raise productivity.

A common feature of critiques of the 1945–51 government, and more broadly of the post-war settlement it embodied, is that the undoubted priority given to full employment damaged productivity. Thus Barnett suggests that there was a direct contradiction between a full employment policy and increased productivity. His evidence for this is, however, non-existent — he refers only to the fact that during wartime full employment productivity was not as high as some unspecified standard that he feels appropriate. Beveridge's view in *Full Employment in a Free Society*, that the end of mass unemployment would remove a major reason for restrictive practices is simply and scornfully dismissed.[7] Barnett's discussion fully justifies Worswick's[8] remark almost forty years ago that 'Full employment itself has become so charged with political and social emotion that a sober assessment of its influence on productivity seems well-nigh impossible'.

Policy-makers in the late 1940s were, unsurprisingly, very conscious of the links between full employment and productivity. What may be called the 'official ideology' of policy-making in this period included the idea that restrictive labour practices were largely a consequence of previous unemployment and insecurity, and that therefore a prime need of economic policy was to convince the workers that full employment was here to stay, and that they could therefore safely abandon such practices.[9]

A strong version of this idea is that a 'bargain' was struck in the 1944 White Paper on Employment Policy, in which full employment was 'offered' by government to the workers in return for raising productivity. In fact, there is practically no warrant for such a view in the White Paper. Only in the last paragraph is there a passing reference to the need to combine productive efficiency with employment.[10] (There is much less of any notion of a contract in this area than in the discussion of labour mobility and wage inflation.[11])

Despite this absence, official trade unionism, in the form both of most union executives and the TUC, strongly committed themselves to the line that restrictive practices were redundant in the new post-war circumstances of full employment.[12]

That union commitment is significant as a symbol of the formal shift in the politics of trade unionism, showing the power of the rhetoric of productivity in this period. It does not, of course, tell us anything about the actual prevalence of restrictive practices, but it did allow a much more open discussion of the issues than seems to have been previously deemed possible.

Study of restrictive practices is beset by severe definitional problems — problems most obviously of drawing lines between 'legitimate' divisions in work practice and those solely imposed to reduce output and increase bargaining power.[13] Leaving on one side these definitional problems, one can examine the evidence produced at the time on what were then perceived as such practices, and ask whether these were related to the full employment prevalent at the time.

One central set of evidence is that from the Anglo-American Council on Productivity. The central preoccupation of most of the reports of the AACP teams was with management, although restrictive practices, broadly defined, are often mentioned. But in only one of my sample of twenty of the forty-seven industry team reports are these practices spelt out.[14] This is in cotton, where the specific restrictive practice cited is 'union resistance to re-deployment', which meant task redesign, involving a greater distinction between the jobs of skilled/unskilled workers.[15] A feature of this not cited in the AACP report was that the main redeployment involved upgrading the skilled *women's* jobs and concentrating the unskilled tasks on male helpers which may well have increased resistance.[16]

The AACP explanation for this resistance to change was summarised in characteristic terms: 'Because memories still remain of the dark days of unemployment, short time, the Means Test and hundreds of other vestiges of the past, trade unionism in Britain is falling short of the responsibility for the future that has been placed before it'.[17]

The only other AACP report where a great deal of emphasis was put on restrictive practices was that on grey iron-founding, but like so many other discussions this was in general terms, the only distinguishing feature being an explicit reference to the idea that 'the British policy of full employment has removed an incentive to effort'.[18]

Overall, the striking feature of the AACP reports is the *limited* role assigned to restrictive practices in accounting for British problems, in a context where such an ascription would have been unusually politically acceptable. (A more general discussion of the AACP is given below).

The government used its productivity drive to pose questions about restrictive practices to employers. A sub-committee of the National Joint Advisory Council (the main tripartite body dealing with industrial relations at the national level) set up a sub-committee, but unfortunately the relevant file remains closed.[19] However, the Ministry of Supply conducted separate discussions with engineering employers, and the Minister reported that 'I have heard only one complaint of lack of co-operation from workers' — the shuttle-makers union insisting that all parts of the shuttle be made by craftsmen.[20]

Zweig's study of *Productivity and Trade Unions* found only one clear case of lack of co-operation by trade unions with the productivity drive — again in cotton. He reported on the 'Nelson campaign' of April–June 1948 which involved calls for both increased labour in the cotton mills and increased labour effort from those already employed. His comment on the latter aspect was:

> I believe it is fair to say, that the majority of the operatives did not respond to the Campaign, and many of them even resented the exhortation, feeling that they were pulling their weight and doing normally what they could.[21]

Thus contemporary enquiry and discussion do not provide good evidence for the thesis of a significant 'restrictivist' response to full employment. The one industry where the issue clearly figured was cotton. This is an interesting case, because not only did this industry have a long history of unemployment and short-time working, but it was well accepted that the boom in demand for the industry's products was likely to be temporary. Once Japan revived, then the long-run decline of the industry was expected to reassert itself.[22] Perhaps cotton workers had rational expectations.

Overall it is hard to find evidence for the Barnett thesis of a negative relation between full employment and productivity.

Turning to other parts of the package of post-war reforms, many of them seem to have had limited short-run effects on the performance of the economy. Of course, welfare state reforms meant higher public expenditure than would otherwise have been the case, but no convincing case has ever been made for an inverse relation between enhanced productivity and economic growth, and public expenditure levels.[32]

Whilst personal taxation in this period was high by historic peacetime standards corporate taxation was only 10 per cent on undistributed profits after 1947, and profits expanded sharply, rising 250 per cent in money terms 1938–48, whilst wages rose 200 per cent. Rates of return in the private sector were around 14 per cent in 1948, again roughly two and a half times those in 1938.[24] There is no evidence that welfare expansion depressed investors' profit expectations. Indeed the evidence is compatible with the opposite case — that the welfare state commitment encouraged private sector investment, as it promised a much more buoyant economic environment than in the 1930s. So it is difficult to argue for 'psychological' crowding-out in this period.

The most specific alleged contradiction between 'welfarism' and growth is the area of investment. The argument is that the wartime and Labour governments committed themselves to a housing programme which crowded-out private, productive investment and thus reduced industrial expansion.[25] Leaving aside for the present the general issue of the role of investment in Britain's productivity problem, how far is it plausible to say such crowding-out took place?

If crowding-out took place in this period it was not in the standard manner of financial crowding-out, with government borrowing bidding up interest rates. The government budget was in surplus from 1947 and interest rates very low. The constraint on investment was via government controls, mainly physical in character. These controls undoubtedly did reduce investment below what it otherwise would have been — plainly there was excess demand for investment goods in this period. However, for most of the Labour government's period in office the competition with industrial investment for physical resources — the most important of which was steel — came not from housing but from exports.

By 1949 steel allocations to housing, education and health combined were down to 215,000 tons out of a total of 6.3m tons. Most of the remainder went to exports and domestic investment.[26] Whilst housing did, of course, absorb resources, by and large these were not those crucial for the scale of industrial investment.

More generally it is worth noting that for all the undoubted political priority attached to housing, especially at the beginning of the Labour government's period in office, the share of total investment going into dwellings under Labour was almost exactly the same as 1925–37, and that since that time the relative costs of housing had risen sharply, and in any event after 1947 the trend in this share was downwards.[27] At a more aggregated level, whilst investment as a share of GNP was lower than, for example, France (9.2 per cent compared with 13.2 per cent in 1946) the rate of increase was proportionately greater in Britain — leading to levels of 13.3 per cent and 17.2 per cent respectively in 1952.[28] The same point is made in a slightly different way by Cairncross when he notes that investment rose by 58 per cent 1945–52, whilst GDP rose only 16 per cent.[29]

A case can be made that investment under Labour was excessively constrained by the emphasis on export expansion (sterling exports rose 77 per cent in the period 1945–51). Such a case might suggest that the balance of payments problem should have been much addressed by cutting overseas commitments rather than by diverting so many resources to exports. But, whilst this is an interesting area of speculation, the case has not yet been persuasively made; those who criticise the scale of Britain's overseas commitments in this period do not link this criticism to investment.[30]

1 Labour and the policy concern with productivity

Labour put a high priority on raising productivity. This priority can be measured in part by the emphasis on productivity in statements on economic policy. Most notably in the *Economic Surveys* of the late 1940s there is a clear progression from the 1947 *Survey*, with its focus on the need to raise production to finance a full employment level of imports, where the constraint on production is manpower, to the 1949 *Survey* where with no more obvious reserves of manpower to draw on, raising labour productivity is the only answer. The 1949 and the following *Survey* in 1950 make productivity *the* central concern.[31]

The evolution of the *Economic Survey* is important, because it brings out how the concern with productivity evolved from the Attlee Government's initial belief (obviously inherited from the war) that manpower planning was the central instrument of resource allocation. A concern with labour productivity was implanted which survived the abandonment of manpower planning and the move to fiscal policy as the dominant instrument of economic policy, because that dominance very much emerged in the context of a supply-constrained economy.

The *Economic Survey* was the most obvious manifestation of shifts in this period. But any reader of the Public Records in this period will see this as merely the tip of the iceberg.

Another important index of government concern was the setting up of a high-powered Committee on Industrial Productivity in 1947, chaired by Sir Henry Tizard. It is clear from the circumstances surrounding the setting up of this committee that raising productivity was very much seen in the light of short-run macro-economic problems (i.e. the balance of payments) and rather ludicrous ideas were entertained about the speed with which productivity expansion could be brought about, hopes which were soon explicitly dashed by the committee.[32]

But whilst it is hardly surprising the macro considerations affected the manner in which the productivity problem was discussed, it would be wrong to think that productivity was only a transient concern, viewed as a short-term 'fix' for a desperately supply-constrained economy. Whilst this was an aspect of the policy framework, there was also a longer time-scale operating in policy discussions where it was accepted that Bitish industry was suffering from long-existing deficiencies which could only be addressed by long-term policy instruments.[33]

At the most public level the concern with productivity was apparent from 1946 in the Production Campaign and the stream of exhortation which flowed after this. This campaign ranged from general media advertising, high level conferences, factory-distributed notices and newspapers.[34] As Cairncross [35] asserts, 'few Governments have proclaimed more insistently the need for higher productivity'. But beyond exhortation what policy instruments were deployed by Labour to raise productivity?

2 Policy instruments

Few economic policy instruments have single conditions of emergence or single objectives, and this would certainly apply in the area of policy on productivity in the immediate post-war period.

Undoubtedly the best-known agency concerned with productivity in this period was the AACP.[36] Strictly speaking this was not a governmental agency. Whilst the initiative came from the European Co-operation Administration head Paul Hoffman (ex-Studebaker), and was enthusiastically taken up by Chancellor of the Exchequer Cripps, the government explicitly saw a political advantage in devolving responsibility for much activity on the productivity front on to a joint employer–union body, which, with the addition of ECA representation, was what the AACP became. In this way the government hoped to involve employers and union representatives more fully in the productivity drive, but to evade any responsibility if the drive failed.[37]

This ploy seems to have been welcomed by both employers and unions as a way of getting the government 'off their backs'. Employers thereafter used the existence of the AACP as a reason for resisting other government productivity initiatives (e.g. in the important area of standardisation).[38]

The most important product of the AACP's endeavours was the production of sixty-six reports on sectors or functions of industry. It is very difficult to summarise these reports, as many had substantial lists of factors affecting productivity rather than a clear message. It is thus easy to select the message of the AACP reports congenial to one's own prejudices. One sustainable assessment of those reports is that they strongly emphasise *managerial* responsibility for the failings of British industry.

The view that the AACP reports were above all a critique of British management seems to have been taken by the Federation of British Industry, whose attitude to the AACP was always lukewarm, and which resented the tone of the reports, and the implicit challenge to managerial legitimacy they contained.[39]

Overall it would seem the government's attempt to use the AACP to engage the enthusiasm of employers and unions for the productivity drive failed. There is little evidence that the AACP reports had much impact except upon the already converted. Employers had little incentive to push the Council's conclusions in the economic context of the time (a point returned to below) and politically the emphasis of most of the AACP reports had little appeal.[40]

The circumstances of the AACP's creation were highly specific, flowing from the Marshall Aid policy, and the consequent British desire to appease US opinion on the performance of the British economy.[41] Less conjunctural in origin were the Development Councils.

Development Councils were tripartite bodies for particular industries, the idea for which had its origins in the rationalisation movement of the 1920s and 1930s. That movement was seen by the Labour Government to have failed to generate adequate capacity to rationalise and improve industrial performance, and what was needed was a new institutional framework to pursue those goals more effectively. This view was reinforced by the perceived success of the wartime concentration of production policy,[42] and by most of the working parties set up by the government to report on a range of mainly consumer goods industries in the late 1940s. Most of these working party reports saw the need for some statutory body to pursue efficiency goals.[43]

Initially in 1945–47 the move to set up Development Councils was not opposed by employers, who saw them as a possible restrictionist shield when the expected post-war slump arrived. But from 1947, with this slump seemingly postponed indefinitely, and the political legitimacy of the Labour government undermined by economic crisis, employers withdrew support from such statutory bodies. The government persevered with the idea, passing an Industrial Development and Organisation Act in 1947 which allowed, as a last resort, the imposition of such Councils even where employers were opposed. Eventually only four were established — in jewellery and silverware, lace, furniture and cotton. Of these, only that in cotton was significant, and this rested on two features. First, the Cotton Development Council to all intents and purposes was the old Cotton Board (created 1940), and its existence reflected a pre-existing consensus on the need for such a body in the industry. Second, the government used subsidies to encourage rationalisation schemes under the aegis of the Cotton Development Council, and only enterprises above a minimum size (measured by number of spindles) qualified for investment grants.[44]

The statutory powers of the other Development Councils did *not* include any capacity to rationalise, but mostly related to the right to a levy to finance research and marketing. The Development Councils at best became an agency for the provision of common services to fragmented industries rather than an agency of rationalisation.

The hostility of employers to Development Councils was grounded partly on a straightforward dislike of a governmental role in industrial affairs (hostility to a role for unions seems much less prevalent). More interestingly, hostility was grounded on the role of Trade Associations, which had been greatly strengthened by the war not only by their general role as mediators between firms and government, but also and more importantly because of their role as administrators of physical controls. When those wartime controls were continued after 1945, Trade Associations became willy-nilly key agencies of economy policy. From this position they were able effectively to organise and articulate opposition to government policies on DCs which they believed threatened their role. Whilst the government pressed on with the IDO Act in the face of this hostility, its persistence was ultimately seen as necessary more to maintain political credibility rather than in the expectation of having a significant impact on industrial organisation.[45]

Even more ambiguous in their status were Joint Production Committees — workshop-based bipartite committees. These had been widespread in the war period, especially in engineering. From 1947 the government strongly encouraged their re-establishment as part of the productivity drive (rather than because of any commitment to 'industrial democracy').[46]

At least in a nominal sense the JPC campaign was quite successful — many such committees were established in the late 1940s. However, they were not welcomed with enthusiasm by employers who feared them as a Trojan Horse for workers' control. Official trade unionism was generally keen, but at workshop level this enthusiasm was less prevalent. Perhaps the workers who elected a deaf-mute to the JPC were not typical, but many seem to have been little more than welfare, 'tea and toilets', discussion shops.[47]

Government ambitions for JPCs were not entirely coherent. On the one hand the driving force behind their encouragement was Cripps. Though he was leader of the productivity drive, his intellectual framework was much influenced by the human relations school of industrial social psychology, particularly via the work of the Tavistock Institute. (This linkage is very apparent in the Tizard Committee's sub-committee on 'Human Factors'.)[48] But this framework was not easily reconcilable with the belief in rapid productivity gains from JPCs, given its emphasis on the necessarily gradual evolution of high trust relations at enterprise level. Partly because of this framework, the calls for the compulsory imposition of JPCs from both unions and the Labour Party Conference were resisted by the government, and increasingly JPCs became incorporated into the Ministry of Labour's strategy for establishing a viable 'voluntarist' framework of industrial relations over the long term, rather than having much directly to do with productivity.[49]

AACP, DCs and JPCs are the trio of high profile instruments of productivity policy in this period. But other policies, perhaps potentially more significant, were also pursued, albeit in a much more diffuse manner. These were in the areas of education and training, including management training, management consultancy, and research and development (R & D).

In school education, Labour was the legatee of the Butler Act. There were growing debates within the party about the 'multilateral' or comprehensive school, although they had little effect on policy. Very much less attention was put on the curriculum or, in particular, the question of technical education at school and post-school levels.[50] The 1944 Act made provision for technical education both through the idea of technical high schools, and by envisaging compulsory day-release education for all up to 18 years of age. But neither of these was pursued with vigour by the LEAs who retained reponsibility.[51]

On post-school education, it was widely discussed and agreed at this time that Britain was deficient in technical education, but there was much less agreement on what to do about this.[52]

Bevin, as Minister of Labour during the war, had put high priority on technical education. Over the 1939–45 period day-release students rose from 42,000 to 150,000, full-time (intermediate) technical college students rose from 13,000 to 31,500, part-timers from 51,000 to 178,000. But the numbers declined rather than increased in the post-war period. The reasons for this seem unclear. The traditional, school-oriented priorities of the Ministry of Education may have reasserted themselves — they 'clamped down on practically every proposal which was submitted to them'. Money for buildings was channelled into schools rather than further education: £208m for the former sector, compared with £21m for the latter over the decade following 1945.[53]

Much more of the time of policy-makers was devoted to discussing higher technical education. In 1938 Britain had produced only about 2,500 graduate or professionally qualified technologists. Here again, there were significant wartime initiatives, but these were not followed through with vigour.[54] The Percy Report of 1945 recommended *inter alia* that National Colleges for specific subjects be founded. This came about by 1950 in the cases of foundry, rubber technology, heating and ventilation, but these colleges were not upgraded to university level as Percy advocated.[55] This status issue was the main area of contention, which prevented much policy progress. Both Percy and the Barlow Committee[56] recommended upgrading some technical colleges to univeristy standard, to provide more technically qualified graduates. But there was no agreement on this, as opposed to the alternative of the universities taking on the task. In the event university expansion did take place, but there was no breakthrough in the numbers of such technical students: numbers roughly doubled 1938–50/51, but then fell as the war-induced backlog of demand for places was satisfied. There continued to be a clear excess demand for graduates in technology and engineering throughout the late 1940s and early 1950s.[57]

On management education, Labour was able to build on the existence of the British Institute of Management, which Dalton as a Coalition Minister had played an important role in creating in wartime. The major function of this was educational, but the Treasury believed it should move rapidly towards self-financing, and hard battles had to be fought to keeps its grant going under Labour.[58]

Most supply-side accounts of British economic failure assume that more management education is an unqualified good thing. But as Armstrong[59] has persuasively argued, what needs to be assessed is also the curriculum of such education. From study of the curriculum of this period he argues that the drive to increase management education assumed that such education should be along the lines of the universalistic and abstract pattern of Harvard Business School, rather than that of the industrial engineering tradition of some American schools. This, along with the human relations approach also popular at the time, pushed British management education away from the area of specific competences — especially in the area of production engineering — where British deficiencies were arguably the greatest.

Governmental provision of firms of management consultancy arose, like the BIM, from the wartime Ministry's realisation of the deficiencies of the British management they had dealings with. A diffuse set of initiatives resulted from this realisation, but the one taken up by the Attlee Government was the Production Efficiency Service, originating in the Ministry of Aircraft Production.[60] The proposal for such a governmental role in raising efficiency in the private sector was resisted by both union and employers' leaders. Citrine for the TUC saw it as raising profit levels by government and potentially causing unemployment, and argued that 'workpeople should not be told how they were to do a job'. Ramsay for the employers said the proposed PES 'seemed to do at Government expense what industry was quite prepared to do for itself.[61]

Nevertheless the government went ahead with creating such a body. However, the scale of activity was small — the PES spent £35,000 in 1948 and dealt with 162 firms' enquiries in the year to 31 May 1949. But its actions were strongly circumscribed by divisions within the Board of Trade over the scope of its activities, as well as by a shortage of trained personnel.[62]

On R & D, British deficiencies in this field were widely recognised in and out of government. A survey by the FBI in 1943 estimated UK expenditure on R & D by private firms at only £5.4 m in 1938, and suggested that 'a wide field exists in which research is not being undertaken at all OR is being undertaken on an inadequate scale'.[63] Individual firms' deficiencies are apparent in the discussions of ICI and Courtaulds at this time.[64] This problem was addressed in a number of ways by the government, but primarily by encouraging Industrial Research Associations via the Department of Scientific and Industrial Research. Total government expenditure on civil research (including that by research councils and universities) rose by almost four times in real terms between 1945–6 and 1950–1. In this area Labour followed a clear bipartisan policy, established during the war, that more resources should be devoted to civil research, especially on the more applied aspects of science. More explicitly a Labour policy was the establishment of the National Research and Development Council to overcome the perceived inadequate private backing of invention.[65]

Labour policy thus reinforced a pattern in which an unusually high proportion of R & D by international standards was financed by public money. But, in addition, the Labour Government originated the position whereby a disproportionate share of British civil R & D was spent on areas of limited commercial impact — notably atomic energy and jet aircraft. Both of these were, of course, interwoven with explicitly military R & D, which was twice as high as publicly funded civil research by 1950–1.[66]

Here, perhaps, was the most clear-cut problem of potential crowding-out of the period — the crowding-out of the supply of labour for commercially significant R & D by a combination of quasi-civil and military research whose rationale lay ultimately in a search for retention of great power status rather than economic advance.[67]

3 A lost opportunity?

The above sketch of the Labour Government's activity on the productivity front clearly gives the lie to any notion that the government was so obsessed with welfare policy and 'New Jerusalems' that it neglected the issues of production and productivity. And, whether caused by government policy or not, labour productivity did rise in this period, the change in trend seeming to come in 1948 to reach about 2.5 per cent for the whole economy and 3.5 per cent for manufacturing over the next three years compared with perhaps 1 per cent and 2 per cent previously.[68] But by the normal standard, compared with other major Western European countries, British productivity was on a path of deficiency. Could the government have done more?

The literature contains a number of views on this. Cairncross robustly defends the macro-economic record of Labour:

> Few Governments also have held back consumption more assiduously so as to let the pace be set by exports and investment as recommended by a later generation of experts on growth. They were successful in achieving a fast growth in exports, eliminating in turn the external deficit and sustaining a high level of industrial investment in spite of the virtual cessation of personal savings. But they did not succeed in raising the rate of growth to the level that their European neighbours proved capable of maintaining. *It must be very doubtful whether any set of Government policies could have done more.*[69]

This scepticism is not justified at length, but is linked to the question of incentives, and the comparison made with France and Germany, where it is suggested redistributive policies were pushed less far and so 'French and German workers consequently addressed themselves to the task of earning an adequate living with a greater inducement to exert themselves'.[70]

The comparison with France is certainly worth pursuing. It is a commonplace to contrast the French emphasis on planning for growth in this period and the British emphasis on what may be called planning for stabilisation. Views on the significance of French planning for French economic growth vary greatly; Cairncross has suggested that 'French planning was largely a bad joke when you look at in detail'. Others have seen a positive role, and lamented the absence of a parallel British system.[71] Whatever its substantive contribution to French economic performance, French planning can surely be said to symbolise the political priority given not only by French governments but French society generally to the raising of economic activity. Monnet linked this to military defeat. 'Britain had not been conquered or invaded; they felt no need to exorcise history'.[72] Whatever the reasons, it is clear that all political forces in France embraced growth and productivity increases in a way which is not apparent in Britain. Sir Alec Cairncross may or may not be right about the attitudes of French and British workers (I know of no decisive evidence on this point), but the contrasting attitude of employers, which is surely more important as it feeds not only into behaviour but determines organisation, investment and all the other crucial decisions for productivity in a capitalist economy, seems apparent. As the Productivity Committee noted, 'the decisions which lead to improvement in industrial productivity, whether in the nationalised or the private sector must be taken by management. Without initiative by, or at least co-operation from, those managements there is little Government departments can achieve'.[73]

Franco-British management differences are nicely illustrated in their responses to the US initiative under Marshall Aid, which led to AACP in Britain and the parallel Association Français pour l'Accroissement de la Productivité (AFAP). Not only were the French missions several times as numerous as the British (450 compared with 66), but the reception in the European countries markedly different. Whilst in Britain employers treated AACP as a way of getting government off their backs, and little enthusiasm was evinced for the 'American Road', in France 'Americanisation' was enthusiastically embraced by much of management.[74]

British employers' unenthusiastic response to government productivity measures can be explained by three factors. First was the lack of compelling economic reason for pursuing enhanced efficiency — this was very much a period of a sellers' market, with producers able to sell anything produced at high profits. Second, there was the obvious political response to a Labour government — a fear that its policies involved an over-extension of government. Third and relatedly, was the belief that government emphasis on efficiency and productivity was an attack on the legitimancy of existing management and employers.[75]

Thus one could from the argument so far suggest a story that, far from Attlee's government pursuing welfarism at the expense of industrial efficiency and competitiveness, it tried very hard to raise industrial growth but was resisted by obdurate employers and management. Such a story some would find ideologically congenial, and it would contain a considerable element of truth. But it would be facile and inadequate because it would not address the issue of how Labour approached the private sector, where of course, most of its productivity policies would have to be effective if they were to succeed.

The Labour government's attitude to the private sector was part of its broader emphasis on consensus and co-operation.[76] Alongside its commitment to nationalisation was a policy on the private sector which combined (at least till 1948) a continuation of physical controls coupled to a large degree of 'self-government' for industry. Indeed in some ways that self-government extended to those controls, which were largely adminstered by Industry and Trade Associations.[77]

The use of inherited physical controls over the economy initially served to hide the almost total absence of a Labour policy on the private sector. Much Labour discussion of the 1940s treats the private sector as that which is left over after nationalisation, and characteristically *Let us Face the Future*, the election manifesto of 1945, had little to say on the private sector beyond advocacy of monopoly policy[78] and a general commitment to industrial tripartism.[79]

The problem was that the physical controls Labour inherited were very much geared-up for dealing with a shortage economy where price rises to equate supply and demand were ruled out. As those shortages diminished as output expanded Labour found little political sense in continuing the controls and little alternative basis for policy in the private sector. In addition those controls had become less popular on the Left as they were seen as strengthening the power of bodies like Trade Associations rather than being clearly agencies of public purpose.[80]

Towards the end of the Labour government, Harold Wilson (President of the Board of Trade 1947–51) in a long and closely argued paper was to lament the inadequacy of Labour's approach to the private sector: 'in this problem of the relation between Government and private industry we have what is almost a vacuum in Socialist thought'.[81] What is most striking about Wilson's paper is that it stands alone as a systematic discussion of Labour and the private sector in this period. It signals the recognition by Labour that the mixed economy is here to stay, and that Labour must fill the vacuum in its policies towards the private sector, which can no longer be regarded as destined eventually to be nationalised. The central thesis of Wilson's paper is that the post-war controls were ill-fitted to the task of raising efficiency, and that what Labour needed to do was to think of policy instruments which would affect decision-making at the Board level of industrial enterprises, where the crucial efficiency-relevant decisions were made.

Whatever the merits of Wilson's paper, it throws into stark relief the void in the whole area of Labour's relation with the private sector, and in particular its lack of attention to questions of industrial managment.

Partly the explanation for this lack is a simple and obvious one, the Labour Government in the late 1940s was faced with an unprecedented situation, where the primary objective of policy was, unexpectedly, not to prevent unemployment and output loss, but to galvanise the economy into producing more from a more or less given labour force and highly constrained capital stock.

Nevertheless, the government's response to this new situation rested on long-standing features of Labour's doctrine and organisation. The central problem was Labour's ignorance of and lack of concern with enterprise functioning and management.[82] This partly related to the character of Labour's social base — predominantly in trade unionism and the non-industrial middle classes.[83] The dominant ideology relating the Labour movement to the private sector was 'voluntarism', a hostility to state intervention in the labour market, which ultimately rested on the notion of management as a hostile social force. This attitude in turn inhibited the development of a social base for Labour in the managerial class to articulate and develop a specifically managerialist ethos in the Labour Party.

It is too facile to link this approach to management solely to Labour's position as a trade union-based party. Even those intellectuals most concerned to develop a distinct democratic socialist (revisionist) position for the Party shared the general unconcern with management. For example in Jay and Durbin's self-consciously revisionist works, the treatment of management is solely in the context of an attempted refutation of Marx's alleged prediction of class polarisation, linked to the emerging sociological theories about a 'managerial revolution'. Pessimism about managements' political predilictions is coupled with an absence of discussion of what management in the private sector actually does.[84]

Of course these are broad generalisations to which exceptions must be admitted. Ministers like Cripps and Dalton were, from their war experience as Ministers, highly concerned with issues of management. MPs like Austen Albu and Ian Mikardo tried to develop a Labour managerialism, in terms of both an ideology and a social base.[85] But even in these cases the predominant view of management was to a striking degree that of the 'human relations' school, with its very narrow conception of the management function, and one perhaps particularly difficult to reconcile with Labour movement traditions.[86] In any event the extent to which these figures remained exceptions is shown by the paucity of discussion of controlling and managing private industry in Labour's wartime reconstruction proposals.[87]

The general weakness of Labour's position in this area is illustrated if we look at the origins of its post-war proposals. Broadly speaking these originated not from within Labour circles, but from the Board of Trade's crucial discussions of 1943[88] *minus* the proposed Industrial Commission *plus* the JPCs *plus* the AACP. The JPC's exception owed much to Cripps' idiosyncratic links with psychology, and was soon smothered in the much more powerful force of industrial relations voluntarism; the AACP, of course, arose from the highly specific shift in US foreign policy, the onset of Marshall Aid, and the British desire to appease American opinion.

Thus Labour in 1945 was a very 'unmanagerialist' party, which along with all the other constraints made it very difficult to play the part of a party of industrial modernisation as well as a party of social reconstruction. Whilst this essay cannot seriously address the issue, it would be instructive to see how deep the apparent change in Labour policy was by the 1960s, when a substantial part of the Party's claim to office was precisely as a moderniser of Britain's industrial apparatus. Was industrial efficiency and productivity grafted more successfully on to the politics of labour in that period than proved possible in the 1940s?

Notes

1 This is a revised version of a paper first given at the Economic History Society Conference in April 1989. I am grateful to Martin Cave, Helen Mercer and the Editors of this book for helpful comments.

2 National Economic Development Office, *British Industrial Performance*, London, 1987, pp. 30–6.

3 A. Cairncross, *Years of Recovery: British Economic Policy 1945–51*, London, 1985; W. Crofts, *Coercion or Persuasion?: Propaganda in Britain after 1945*, London, 1989, chapter 5.

4 K. Middlemas, *Power, Competition and the State, Volume I: Britain in Search of Balance*, London, 1980.

5 Cairncross, *Years of Recovery*, chapters 1, 2, 4.

6 R. Jones, *Wages and Employment Policy 1936–1985*, London, 1987, chapter 4.

7 C. Barnett, *Audit of War*, London, 1986, pp. 258–63, 262–3.

8 Worswick, 'Introduction', in G. D. N. Worswick and P. H. Ady (eds), *The British Economy 1945–70*, Oxford, 1952, p. 8.

9 A 'classic' statement of this is by the TUC *Production Under Full Employment*, PRO LAB 10/655, JCC 183, 5 November 1946. See also CAB 134/187, Official Steering Committee on Economic Development, *Minutes*, 15 October 1946.

10 Cmnd. 6527, para. 87.

11 Cmnd. 6527, paras, 56 and 49 respectively.

12 TUC, *Trade Unions and Productivity*, London, 1950.

13 F. Zweig, *Productivity and Trade Unions*, Oxford, 1951.

14 This is all the more striking given the management bias in the composition of the AACP teams. See A. Carew, *Labour Under the Marshall Plan*, Manchester, 1987, pp. 138–9. There were sixty-ix team reports in all. Nineteen were specialist reports, e.g. Education for Management, Universities and Industry, Plant Maintenance. Three of the forty-seven team reports were of US visits to Britain, in cotton, the electricity system and pressed metal. (These figures are derived from the list of reports in the *British Cotton Industry Report*, London, 1952. These figures are slightly different from those in AACP *Final Report*, London, 1952.)

15 AACP, US Team Report on the *British Cotton Industry*, pp. 22–3.

16 PRO, LAB 10/656, Joint Consultation Committee Papers, *Cotton Industry* (no date, but 1947).

17 AACP, *British Cotton Industry*, p. 24.

18 AACP, *Grey Iron Founding*, London, 1950, p. 14.

19 PRO LAB 10/658, closed to 1991.

20 PRO CAB 134/645, Production Committee, *Discussion with Engineering Industry on Dollar Drive and Productivity*, 1949.

21 Zweig, *Productivity*, p. 159; see also Crofts *Coercion* chapters 8–10.

22 PRO LAB 10/656, JCC, *Cotton Industry*; J. Singleton, 'Planning for cotton, 1945–51', *Economic History Review*, XLIII, 1990, pp. 62–78.

23 D. Heald, *Public Expenditure: Its Defence and Reform*, Oxford, 1983; S. Brittan, 'How British is the British sickness?', *Journal of Law and Economics*, XXI, 1978, pp. 251–3.

24 T. Barna, 'Those "Frightfully High" Profits', *Bull. Oxford Institute of Statistics* II, 1949, pp. 213–26; A. D. Rogow and P. Shore, *The Labour Government and British Industry*, Oxford, 1955, pp. 68–9.

25 Barnett, *Audit of War*, pp. 242–7; M. Chick, *Economic Planning Managerial Decision-Making and the Role of Fixed Capital Investment in the Economic Recovery of the U.K. 1945–55*, unpublished Ph.D. thesis, University of London, 1986, pp. 38–43. The discussion here follows J. Tomlinson, 'Labour's management of the national economy 1945–51', *Economy and Society*, XVIII, 1989, pp. 1–24 and J. Tomlinson, *Public Policy and the Economy Since 1900*, Oxford, 1990, chapter 8.

26 PRO CAB 134/191, Official Committee on Economic Development, *Minutes*, 20 July 1949.

27 R. C. O. Matthews, C. H. Feinstein and J. Odling-Smee, *British Economic Growth 1856–1973*, Oxford, 1982 pp. 332, 413–14; Cairncross, *Years of Recovery*, p. 456. N. Tiratsoo, *The Reconstruction of Coventry* (forthcoming), ably brings out the subordination of housing to industrial expansion in Coventry in this period, and Coventry was a case where the political priority of housing was particularly compelling.

28 Chick, *Economic Planning*, chapter 1.

29 Cairncross, *Years of Recovery*, p. 24.

30 E.g. E. Brett, S. Gilliat and A. Pople, 'Planned trade, Labour Party policy and U.S. intervention: the success and failure of postwar reconstruction', *History Workshop*, XIII, 1982, pp. 130–42.

31 Economic Surveys for 1947–50.

32 The Committee issued two reports. 1st Report Cmd. 7665, 2nd Cmd. 7991. The proceedings of the Committee are in PRO CAB 124/1093–7 and CAB 132/28–30. It spawned various sub-committees, recorded in CAB 132/31–50. Also of importance to the productivity debate in the government is the Productivity (Official) Committee CAB 134/591–3 and the Official Steering Committee on Economic Development CAB 134/186–193. The main Ministerial Committee where productivity figured as an important issue was the Production Committee (established October 1947) CAB 134/635–652.

33 E.g. PRO CAB 124/1096, Committee on Industrial Productivity, *Progress Report*, June 1948.

34 S. W. Crofts, 'The Attlee Government's economic information propaganda', *Journal of Contemporary History*, XXI, 1986, pp. 453–71.

35 Cairncross, *Years of Recovery*, p. 499.

36 For accounts of this see A. Carew, *Labour Under the Marshall Plan*, London, 1987, chapter 9; G. Hutton, *We Too Can Prosper*, London, 1953; J. Tomlinson, 'The failure of the Anglo-American Council on Productivity', *Business History* (forthcoming); Leyland, *Productivity*.

37 PRO CAB 134/592, Productivity (Official) Committee, *Minutes*, 30 September 1949; BT 195/19 Committee on Industrial Productivity: Publicity Drive, 15 June 1949.

38 PRO BT 195/19, 19 August 1949, 27 August 1949; BT 195/66 AACP, *First Report and Subsequent Activities*, November 1948.

39 Hutton, *We Too*, chapter 2; K. Middlemas, *Power Competition and the State*, Volume I: Britain in Search of Balance, London, 1980, pp. 162–3.

40 PRO BT 195/66 AACP First Report on Subsequent Activities, *Minutes*, 25 November 1949; BT 195/53 European Co-operation Administration: Technical Assistance, *Minutes*, 12 September 1949.

41 On the US background see C. Maier, *In Search of Stability*, Cambridge, 1987, and Carew, *Labour*, chapters 1, 3–4.

42 On this concentration policy see E. J. Hargreaves and M. Gowing, *Civil Industry and Trade*, London, 1952.

43 P. D. Henderson, 'Development Councils', in Worswick and Ady, *British Economy*, R. S. Edwards and H. Townsend, *Business Enterprise: Its Growth and Organisation*, London, 1962, pp. 386–91.

44 Henderson, 'Development Councils'.

45 PRO CAB 134/637 Production Committee, *Development Councils Under the Industrial Organisation and Development Act 1947*, 19 March 1948. For employers views on DCs see N. Kipping, *Summing Up*, London, 1972, pp. 16–17, 85, 87.

46 For the war period see ILO, *Joint Production Machinery*, Geneva, 1944; P. Inman, *Labour in the Munitions Industries*, London, 1957; J. Hinton, 'Coventry communism: a study of factory politics in the second world', *History Workshop*, X, 1980, pp. 90–118, R. Croucher, *Engineers at War*, London, 1982; A. Exell, 'Morris motors in the 1940s', *History Workshop*, IX, 1980, pp. 90–114.

47 The public records on JPC's in the late 1940s are scattered through LAB 10 and in BT 168/166–173. Also J. Tomlinson, 'Industrial democracy and the Labour Government 1945–51', *Brunel University Discussion Paper*, Uxbridge, 1987.

48 PRO CAB 124/1093, Committee on Industrial Productivity, *Problems Affecting Industrial Morale and Efficiency*, 5 November 1947. On the background see P. Miller, 'Psychotherapy of work and unemployment', in P. Miller and N. Rose (eds), *The Power of Psychiatry*, Cambridge, 1986. The discussions of the Human Factors Panel are at PRO CAB 132/31–35.

49 PRO LAB 10/721, Machinery for Joint Consultation, 26 January 1949; LAB 10/772, JPCs *L. Roberts to Regional Controllers*, 3 November 1948.

50 R. Barker, *Education and Politics*, Oxford, 1972, chapter 5.

51 P. H. Gosden, *Education in the Second World War*, London, 1976, chapter 17.

52 PEP, 'Technological Education', *Planning*, XVII, 1950, pp. 61–80; Gosden, *Education*, chapter 17; G. A. N. Lowndes, *Silent Social Revolution*, London, 1965, chapters 18, 19; AACP, *Universities and Industry*, London, 1951.

53 Lowndes, *Revolution*, pp. 326, 327.

54 N. J. Vig, *Science and Technology in British Politics* 1968, p. 9; H. M. D. Parker, *Manpower: A Study of Wartime Policy and Administration*, London, 1957, pp. 326–30; Barnett, *Audit*, pp. 284–91.

55 Percy, *Committee on Higher Technical Education*; Lowndes, *Revolution*, chapter 19. The Percy Committee was set up to allay criticism about the relative neglect of technical education in Butler's Act.

56 Barlow, *Committee on Scientific Manpower*.

57 M. Sanderson, *The Universities and British Industry 1850–1970*, London, 1972, chapter 7.

58 PRO T228/624–8, British Institute of Management; CAB 124/827, Board of Trade: *Proposals for Improving Industrial Management.*

59 P. Armstrong, *The Abandonment of Productive Intervention in Management Teaching Syllabi*, University of Warwick, 1987.

60 For wartime discussions and policies see PRO T228/624.

61 PRO BT190/2, National Production Advisory Council for Industry, *Minutes*, 7 December 1945.

62 CAB 134/642, First Report of Productivity (Official) Committee, 20 September 1949.

63 Cited in Vig, *Science and Technology*, p. 10.

64 E. g. W. J. Reader, *Imperial Chemical Industries: A History*, Volume 2, London, 1975, pp. 303–5, 447–58; D. C. Coleman, *Courtaulds: An Economic and Social History*, Volume 3, Oxford, 1980, pp. 41–2, 50.

65 Vig, *Science and Technology*, pp. 14–17; Edwards and Townsend, *Business Enterprise*, pp. 372–3.

66 Peck, *Science*; and Vig, *Science and Technology* p. 16.

67 On nuclear power see M. Gowing, *Independence and Deterrence: Britain and Atomic Energy 1945–52, Volume I: Policy-Making*, London, 1974.

68 Cairncross, *Years of Recovery*, pp. 18–19.

69 Cairncross, p. 500, emphasis added. For a similar assessment G. C. Allen, *British Industry and Economic Policy*, London, 1979, p. 114.

70 Cairncross, *Years of Recovery*, p. 504.

71 Cairncross, Letter to Author, February 1989. Compare J. Leruez, *Economic Planning and Politics in Britain*, London, 1975, which assesses British planning from the perspective of the French approach, and P. Hall, *Governing the Economy*, Cambridge, 1986, and J. Zysman, *Governments, Markets and Growth*, Oxford, 1983, which draws a similarly unfavourable comparison. A very early discussion of French planning is by James Meade, PRO CAB 134/190, Steering Committee: *Monnet Plan Anglo-French Discussions in Paris*, January 1947.

72 Cited A. Bullock, *Ernest Bevin, Foreign Secretary*, Oxford, 1983, p. 778.

73 PRO CAB 134/642, First Report of Productivity (Official) Committee, para. 12.

74 L. Boltanski, *The Making of a Class*, Cambridge, 1986, chapter 2.

75 A view supported by some 'pro-market' economists, e.g. J. Jewkes, 'Is British industry inefficient?', *Manchester School*, XIV, 1946, pp. 1–16.

76 Middlemas, *Power, Competition*, pp. 115–16.

77 Rogow and Shore, *Labour Government*, chapter 4.

78 See Helen Mercer in this volume and Mercer, *The Evolution of British Government Policy towards Competition in Private Industry, 1940–1956*, unpublished Ph.D. dissertation, University of London, 1989.

79 K. O. Morgan, *Labour in Power 1945–1951*, Oxford, 1984, pp. 128–30.

80 Rogow and Shore, *Labour Government*, chapters 3, 9.

81 PRO PREM 8/1183, H. Wilson, *The State and Private Industry*, 4 May 1950. See also Middlemas, *Power, Competition*, chapter 6; R. Brady, *Crisis in Britain*, Berkeley, 1950, pp. 551–60.

82 J. Tomlinson, *The Unequal Struggle: British Socialism and the Capitalist Enterprise*, London, 1982; K. Hotten, *The Labour Party and the Enterprise*, unpublished Ph.D. dissertation, London, 1988.

83 B. Hindess, *Decline of Working Class Politics*, London, 1971, chapter 1. Compare C. Newton and D. Porter's *Modenisation Frustrated*, London, 1988, chapter 4 notion of a 'producers alliance' in this period.

84 D. Jay, *The Socialist Case*, London, 1937, chapter 3; E. Durbin, *The Politics of Democratic Socialism*, London, 1940, Part II. Durbin's unpublished work shows a greater concern with management issues, but mainly in the typical economists' concern with incentives. E. Durbin, *New Jerusalems: The Labour Party and the Economics of Democratic Socialism*, London, 1985, pp. 274–6. Carew's *Labour*, pp. 240–2, seems to exaggerate substantially the 'managerialism' of Labour, certainly in the late 1940s.

85 A. Albu, *Management in Transition*, London, 1942; I. Mikardo, 'Trade unions under full employment economy', in R. H. S. Crossman (ed.) *New Fabian Essays*, London, 1953.

86 J. Child, *British Management Thought*, London, 1969, pp. 135–6.

87 I. Taylor, *War and the Development of Labour's Domestic Programme 1939–45*, unpublished Ph.D. dissertation, London, 178, chapters 1, 4.

88 PRO CAB 87/63, Committee on Post-war Employment; Board of Trade, *General Support of Trade*, 15 October 1943. This document also proposed a new financial institution to bridge the finance gap for small firms. This was pre-empted by the creation of FCI and ICFC by the Bank of England. Apart from the nationalisation of the Bank of England, Labour's reforms of finance were strikingly limited: Middlemas, *Power, Competition*, chapter 3; also, A. Howson, '"Socialist monetary" policy: monetary thought in the Labour Party in the 1940s', *History of Political Economy*, XX, 1988, pp. 543–64.

4 *Martin Chick*

Competition, competitiveness and nationalisation, 1945–51

At first glance, it may seem strange to include a discussion of the Attlee government's nationalisation programme in a collection of essays concerned with the issue of competitiveness. After all, it was their very freedom from competition which was the most striking feature of the industries nationalised by the Attlee government. Even accepting the state and municipal involvement in many of these industries prior to nationalisation, the act of nationalisation did effect the consolidation of many previously independent units into massive national single entities. On vesting day, the 3,766 independent management units in the road haulage industry became one single unit, as did the 1,000 in gas, the 560 in electricity and the 800 in coal.[1] Competition within each industry, whether actual or potential, ceased. In addition, competition between industries within sectors like fuel and power, and transport and communication, was replaced by administrative 'co-ordination'. The expunging of domestic competition within industries and sectors by design coincided with the reduction of competition from abroad. Post-war domestic material shortages, balance of payments considerations, restrictions on dollar imports, and the application of price controls to basic domestic output all combined to reduce the level both of imports and exports of coal, for example, thereby reducing that industry's exposure to foreign competition.[2]

The semantic paradox at the heart of nationalisation (or socialisation as it was more commonly termed in the 1940s) was that the reduction of competition within industries and sectors was a response to the increasing concern to improve the 'competitiveness' of many of these industries. The concern with 'competitiveness' had its origins in the inter-war period when difficulties in competing in tight international markets and the consequent high regional unemployment in industries like coalmining, were ascribed to 'uncompetitiveness' and 'technical inefficiency'. Confronted by increasing dissatisfaction with the inter-war performance of many basic sector industries and by increasing concern at the consequences of 'inefficiency' for regional unemployment, the state began to become involved in major industries. At one level, it intervened in the coal industry to mitigate the worst effects of unemployment, to provide subsidies and to slow up the rate of pit closures.[3] At another level, it began to be drawn into the arguments on improving the efficiency of these industries. State intervention to improve industrial efficiency was not new in the 1930s. Following state involvement in the organisation of industry during the First World War, the state at both central and municipal level had become increasingly involved in rationalising and standardising such industries as railways.[4] Concern with the sub-optimal performance of the scattered, multiplant electricity industry and, perhaps most crucially, concern with the high electricity prices being charged, prompted the Conservative Baldwin government to intervene in the electricity industry and to establish the Central Electricity Board and early grid system in 1926. In time, the line between, on the one hand, government concern for the economic causes of inefficiency and on the other, its concern for the political and social consequences of that inefficiency became blurred, the obfuscation becoming greater as government concern with efficiency and, in particular, unemployment widened. The general widening of concern, from specific-industry local unemployment to unemployment within the economy as a whole, was accompanied by a wider and increasingly catholic reading of such terms as 'economic efficiency'. As was noted within the Labour Party during discussions in April 1944 on future employment policy:

> There is more work to do than can be done. There is full scope for the employment of all our labour, machinery and productive power as far ahead as the human mind can see; and the acid test of economic efficiency, for any industrial system or Party or Government, lies in its ability to give these human needs a continuous and effective expression, to provide an ever-widening market for the products of human skill and scientific invention.[5]

In general, it was thought that nationalisation could contribute to the passing of any such 'acid test' by increasing the government's control over the economy and widening the scope for management of the economy. Control of fixed capital investment, in particular, was held to be particularly important, given its potential ability to allow governments to boost or diminish the rate of public investment as the economy slackened or boomed respectively.[6] These ideas, which were to be found in post-First World War Liberal publications, were given added vigour and renewed freshness, not only by the central position given to fixed capital investment activity in Keynes' *General Theory*, but also by the role implicitly available to the state in manipulating investment in order to influence the level of effective demand within the economy. As James Meade, D. N. Chester, J. M. Fleming and others in the Economic Section of the Attlee government[7] hoped:

The socialisation of some industries, can materially assist the Government's achievement of a stable volume of investment as part of the task of maintaining national expenditure.[8]

The exercise of such control was predicated on the assumption that ownership of heavy, lumpy capital-intensive, basic sector industries would be transferred into public hands. The transfer of ownership was thought likely to improve internal incentives to efficiency and, most crucially, facilitate the rapid achievement of economies of scale within the industry. Among the benefits for internal efficiency associated with a change of ownership were improved incentives for workers who would work harder 'because they know the community will be the ultimate beneficiary and not just the shareholders'[9] and an improved quality of management, since managers would be chosen on grounds of 'competence rather than nepotism'[10] More significantly, not only would public ownership provide the funds required for the modernisation of industries like railways and coal, funding which private investors may have been unwilling to provide in full, but the transfer of ownership would also remove the obstacles to industrial restructuring and the exploitation of latent economies of scale which were held to have been inherent in private ownership. This final anticipated benefit, the exploitation of potential economies of scale, was probably the main technical consideration firing the thoughts of aspirant nationalisers.[11] Both the inter-war and wartime experience had convinced many that if industries like coal were allowed to remain in private ownership, then, as in wartime, any proposals for 'drastic action' would be opposed since it 'cut across the boundaries of so many personal interests ... that only half-hearted compromises can be slowly effected.'[12] Extrapolating from the general judgement of many economists and industrial consultants[13] that too many industries in Britain contained too many units operating on too small a scale and 'with a lack of co-ordination which has harmful economic results', the conviction within the Labour Party was that:

Amalgamation and concentration is frequently impossible with the necessary speed unless the State applies compulsion and hence the only effect is to strengthen monopoly.[14]

Not only was it intended that there should be considerable rationalisation within the nationalised industries, but it was also envisaged that these industries would then use their purchasing power and position to promote further standardisation within the economy.[15] The extent of such transfer, whether half, three-quarters or the whole of an industry should be so transferred, was not discussed as a central issue. To an extent, once the need for a transfer of ownership in basic sector industries had been accepted, then, given the emphasis on controlling fixed capital investment projects, any transfer of ownership in heavy, capital-intensive industries could never be on a small-scale basis. Moreover, given the increasing capital-intensity of many basic sector industries, the boundaries of the 'natural monopoly' component within such industries were often widely drawn. The first printed draft of the bill to nationalise the electricity industry provided for a statutory monopoly in both generation and supply. Although references to exclusive rights for the nationalised generators were deleted from later drafts following arguments within the Ministry of Fuel and Power, an internal minute expressed the general view that it was 'most unlikely that private generating plants could be constructed of a size capable of competing seriously with the nationalised industry or that any existing plants could so compete having regard to the great advantages to be gained from public ownership and co-ordination'.[16] Moreover, such nationalised monopoly groupings seemed to offer scope for reducing the level of uncertainty which was perceived as having promoted collusive arrangements between private producers and inhibited more adventurous fixed capital investment decision-making.[17] The almost inherent tendency of lumpy, capital-intensive industries to move towards collusion and risk aversion underpinned much of the Labour Party's thinking on nationalisation. If monopoly or oligopoly was inevitable, then better it be in public hands. As was argued in Labour Party documents advocating the nationalisation of the steel industry:

> it is important to recognise that monopoly is inevitable in the steel industry. Nowhere in the world has it remained in the hands of free competitive private enterprise. And the same reasons have everywhere forced the growth of monopoly. Plant is so big that, using modern methods, a few plants can fulfil most of the demand. New plant is enormously expensive. In these conditions it would be idle to expect competition; the steel firms stand to lose too much. So they agree, by price agreement, quota and restriction, to share the market between them. A further cause of monopoly is that investment and production in steel must be closely balanced with demand, and this can only be done if both are centrally controlled.
>
> Under capitalism, or socialism, central control is required.[18]

Any concern that might have existed at the state establishment of industrial monopolies was tempered by the view that 'free markets' already and increasingly contained informal monopolies. The Labour Party did not regard the abandonment of competition and the 'free market' system as major losses, since the 'free market system' frequently did not operate as its supporters claimed:

> Those who advocate free enterprise ... rest their case upon the theory of competition. Competition, they say, exerts a continuous downward pressure upon prices, so that only the most efficient and up-to-date firms can survive, while any abnormal profits appearing because of a temporary advantage gained by the one firm are quickly wiped out by the other firms following suit. [The Labour Party dispute this.] Even when there is no collusion between firms, competition is very slow and imperfect. The range of costs in the same industry is often staggeringly large, yet the inefficient are not ruthlessly driven out. This is to a great extent due to the lack of standardisation which enables each firm to establish a limited monopoly in a brand

> of a particular commodity. Advertising is widely used to strengthen this monopolistic element, by stressing the differences, often almost imaginary, between one brand and another. In many industries there is tacit, or open, agreement between firms to refrain from price competition, while in some cases we witness full cartel agreements to share the market and to fix minimum prices.
>
> The consequences of imperfect competition and cartelisation are that there is a wastage of productive capacity. Costs are higher than they need be with present technical knowledge. The tacit or open collusion between firms enables them also to reap monopolistic profits, i.e. profits which exceed the level necessary for them to carry on and extend their businesses.[19]

Thus it was that a government which was pushing through the 1948 Monopoly and Restrictive Practices Act was at the same time creating monopolies in many of the basic sectors of the economy. It was not that monopolies in the basic sector were somehow different. Rather the argument was that: 'Monopolies are objectionable and should, other things being equal, be transferred to public ownership'.[20]

It was the view that there was an increasing tendency towards monopoly within capitalist development, particularly in heavy capital-intensive sectors, which made the state control of monopoly increasingly inevitable. The existence of monopolies need not necessarily result in inefficiency. The attention of the Monopolies Commission, established by the 1948 Monopoly and Restrictive Practices Act, was directed towards monopolistic trade practices rather than the concentration of economic power as such. It was the abuse, rather than the existence of monopoly power, which was the principal concern. Having established publicly-owned monopolies, the Labour government was faced with the task of devising incentives and structures to prevent these enterprises abusing their monopoly position. Such abuse did not simply encompass monopoly pricing but a whole range of inefficiencies that might or might not have been tolerated in a more competitive market. Such inefficiencies might result in managers failing to exploit the economies of scale that were held to be facilitated by nationalisation.

Whether available economies of scale were as prevalent as was often suggested was open to doubt.[21] Whether they were or not, there was no inherent reason why the pursuit of improved efficiency and structural reorganisation should have resulted in the adoption of strongly centralised organisational structures in many of the nationalised industries. While contemporary expert opinion emphasised the desirability of a move towards larger operational units, none of the committees of inquiry[22] had recommended units that were national in scale or had visualised nationalisation on the pattern ultimately adopted, namely with a single authority owning the entire assets of the industry. Reorganisation had been seen in terms of compulsory mergers, with area or regional boards as the most obvious alternative to the enlargement of existing undertakings.[23]

In his detailed study of the Attlee government's nationalisation programme,

D. N. Chester felt it 'pertinent to ask why the extension of public ownership and the reduction of the private sector should have been interpreted in Whitehall as meaning the organisation of the industries either on a national scale or in very large regional units.'[24] That the dangers of overcentralised large bureaucracies were understood seems clear. Gaitskell thought it inevitable that compared with small-scale businesses, such organisations would be 'slow, cumbersome, impersonal and probably slightly conservative'.[25] Gaitskell and civil servants within the Ministry of Fuel and Power were keen that the Area Boards (electricity) should be given as large a measure of financial independence as possible and encouraged to pay their way. Property would be vested in the separate Boards, and each Board would contribute to a central reserve guarantee fund, which would provide for any Area Board unable to service its own Stock. Providing a Board paid its way, it would be left to manage its own affairs free from central interference. The appeal of such an industry structure was that it would tend to encourage a form of competition between Boards,[26] as well as introducing efficiency incentives, accountability and decision-making at a reasonably local level. The Treasury also viewed the devolving of financial responsibility on to each Board as providing an important check to excessive development, particularly in the case of such large capital resource users as the electricity industry.[27] Such decentralisation was sought for both the coal[28] and electricity industries, but it was only after the highly centralised structure of these nationalised industries had attracted such criticism, that a greater measure of decentralisation was permitted in the gas industry. Gaitskell made it clear in discussions on the future organisation of the gas industry, that he would 'deprecate more than the minimum of centralisation'.[29] Even here, however, Gaitskell was prevented from achieving as large a measure of decentralisation as he would have liked.

Clearly there were some functions which were better exercised from the centre than from disparate localities. Even the more decentralised structure of the gas industry placed such functions as the organisation and supervision of training and research and development in central hands. Economists like James Meade also saw considerable advantages being derived from the central pooling of relevant information within the industry, the availability of such information being seen as an important step towards reducing uncertainty, particularly as Keynes had argued, as it affected investment decisions. This, after all, had been one of the arguments for nationalisation in the first place. As James Meade argued:

> the competitive market system is less efficient as an instrument for co-ordinating economic decisions affecting the more distant future than for those affecting the immediate future. The advantages of central planning by an industry as a whole are therefore greatest where investment is concerned, and industries which incur a great deal of capital outlay are relatively suitable for centralised control, whether public or private.[30]

Yet, as Meade[31] would probably have been the first to argue, the centralisation of some functions did not exclude the decentralisation of much more information, incentive structures and decision-making.

The principal rock on which the decentralising ambitions of Gaitskell and others foundered was the clear and determined preference of the Lord President, Herbert Morrison, for highly centralised nationalised industries. In the battles over the future organisation of the nationalised industries, such as that between the Ministry of Fuel and Power and the Lord President over the degree of autonomy to be accorded to the electricity distribution boards, the argument was almost always resolved in favour of the more centralised structure.[32] In the dispute over the degree of independence of regional electricity distribution boards Morrison frequently argued that the capital raising and capital servicing requirements for each industry made their centralisation necessary. Certainly, the Treasury was unhappy at suggestions from local authorities that the raising and servicing of capital should be decentralised in a structure containing thirty or more regional bodies because of the sheer multiplicity of Stocks, but it was prepared to countenance such decentralisation in a structure founded on a dozen or so Regional Boards. Again, while it was clearly easier to raise a single Stock for each industry than to handle a series of Area Stocks, this could have been made compatible with the existence of autonomous Boards by making each Board responsible for servicing that part of the total capital of the industry represented by its share of the total assets.[33] It was Morrison, more than anyone else, who pushed for a centralised industry under a National Board. It had the appeal of bureaucratic tidiness, it accorded with notions of central 'planning' and 'co-ordination', and it made the transfer of assets on vesting day much easier. Centralised industries suited the powerful trade unions who wanted nationally negotiated wages and conditions and Morrison thought that centralisation would also make it easier to confer such benefits as cheap electricity on those living in thinly populated areas. Moreover, all of Morrison's previous writing and thinking on the organisation of nationalised industries had concentrated overwhelmingly on the role of the public corporation,[34] an implicitly centralist approach.

The scope for decentralisation and competition within nationalised industries had not been seriously or extensively considered. In part, this was because of the general tendency to view competition as wasteful and exploitative. When economists within the civil service suggested making greater use of market mechanisms in the management of the nationalised industries, their ideas were angrily dismissed by Morrison as being inimical to nationalisation. Moreover, industries like coal into which some degree of competition might have been most easily introduced, were characterised by distorted cost structures which made competition based on comparative performance effectively impracticable. The natural cost differences between coal districts had been accentuated by the introduction of a national minimum wage during the Second World War and by the cross-subsidisation of high cost districts by low cost districts which was effected through the Coal Charges Account.[35] That Morrison was able to establish newly nationalised industries on a more centralised basis than some commentators wished was also a reflection of the contemporary uncertainty about how nationalised industries should be organised and about how they would perform in practice. Some might favour more decentralised organisational structures, but as Gaitskell himself informed the Labour Party's Sub-Committee for Information:

> We shall not really know whether large scale organisation under public control can in every case be made to work efficiently for some considerable time.[36]

and that

> people talk a lot about decentralisation without realising that, to some extent, it may not really be compatible with all the underlying principles and objectives of nationalisation, and, in particular, with the conception of control by Ministers and Parliament over the Boards.[37]

In the rush to push the nationalising legislation through parliament, major issues such as organisation were often left to be worked out later in practice. The task of operating nationalised, largely centralised monopolies efficiently was not addressed adequately at the time of their establishment, this task often being left by default to the civil servants, ministers and managers who would have to run and supervise these industries.

In the early post-Second World War world of excess demand and, by definition, shortages of resources, the efficient use and allocation of resources became of prime concern to economists within government. Charged with such fundamental allocative tasks as 'co-ordinating' the fuel and power, and transport and communication sectors, the achievement of such a task was not eased by the role allotted to pricing within the nationalised industries. The main statutory function accorded to nationalised industry prices was to enable the industry to cover its costs, 'taking one year with another'. The earning of high book profits by nationalised monopolies did not appear to be politically acceptable, especially as one of Morrison's criteria of efficiency was the ability of nationalised industries to provide 'basic output' as cheaply as possible. However, the term 'taking one year with another' did seem to acknowledge that industries were subject to cyclical ups and downs and the implied logic seemed to be that profits could be earned and reserves built up in good years in order to carry the industry through the bad years. Whether this would prove politically acceptable, particularly when the 'good years' might be years of high demand and government worries about inflation, seemed doubtful. Nor was it clear what would happen to industries which suffered a sustained run of bad years and were simply unable to cover their costs over a number of years. In addition, the general injunction that industries need simply cover costs not only removed the profit incentive from managers, but also eliminated one of the yardsticks by which managerial performance might be evaluated.[38] It also implied that prices should be cost-based, but whether this was appropriate and what precisely constituted costs was unclear. For example, simply to allow prices to cover the financial costs of an industry's borrowings from the government would inadequately reflect the resource costs of the industry, something that was of particular importance if co-ordination within major sectors like fuel and power and transport and communications was being sought.

Economists like James Meade in the Economic Section and Philip Chantler in the Ministry of Fuel and Power were particularly keen that prices should reflect the resource cost of the fixed capital investment programmes being undertaken by particular industries.[39] However, differences then arose over whether marginal cost or average cost pricing should be followed, especially if this meant that total receipts were not likely to cover total costs. This was most likely to occur in industries where technological improvements and economies of scale were likely to bring increasing returns on plant. As has been seen, these increasingly capital-based industries were at the heart of many of the arguments advanced by the advocates of nationalisation. In such industries, where average costs fell with increases in output, and in which, consequently, average costs were greater than marginal costs, the proceeds from paying a price equal to marginal costs would be less than total costs.[40] Moreover, as Coase[41] noted, marginal cost pricing carried the risk of benefiting the marginal consumer at the expense of the rest. One possible solution to certain of the marginal cost disputes was the introduction where appropriate of two-part tariffs. Clearly in industries like electricity where peak-hour demand was determining capacity requirements, there was much to be said for such tariffs. Yet, there was considerable opposition to their implementation from such industries as electricity. Deprived of profit incentives, but firmly driven by expansionary, sales-maximising ambitions, the industry used its industrial political strength to obstruct efforts to introduce effective tariffs and to make even the seasonal differential introduced by Gaitskell following the recommendations of the Clow Committee[42] too small to be effective. Although Gaitskell's proposed increase in winter electricity prices was opposed fundamentally because of its potentially dampening effects on the electricity industry's sales-led programme of expansion, the industry was able to exploit a serious practical problem in many of the arguments for (marginal) cost-based pricing. Gaitskell, Clow, Chantler and others supported the case for marginal cost pricing in part because they saw it as offering, in the short term, one chance of beginning to get to grips with the problem of excess demand, and, in the long term, because they saw it as the most efficient basis for resource allocation. However, the argument made by the electricity managers was that cost-based price increases were largely irrelevant to solving the excess, peak-hour, demand problem, for which they proposed supply-side, plant expansion, solutions. No one knew what the price elasticity of peak-hour demand was likely to be but it seemed very unlikely that cost-based pricing, as opposed to demand-based prices, would be effective in choking back demand. Indeed, many of the early post-war discussions of average/marginal cost pricing look curiously academic in an economy in which the control of excess demand was the dominant problem. To be fair, economists did recognise that the immediate post-war period was a difficult time in which to establish long-term cost and pricing policies. As the Economic Section noted:

The following observations on price policy in socialised industries are intended to present an ideal, applicable to conditions of comparative normality, towards which we should steadily move, rather than something to be put in its entirety, into immediate operation. ... Moreover, some of the principles advocated here are not strictly applicable to conditions of shortage and potential inflation when the pricing system is necessarily in eclipse, and some of its functions are for the time being exercised by controls of the war-time variety. But the pricing problems of the transition period will be more happily solved if the longer-term goal is kept in mind'.[43]

Nevertheless, a contemporary criticism frequently levelled at economists was that while they developed a set of principles on which the nationalised industries might operate in the long term, they ignored the practicalities of the immediate present. Moreover, economists were also criticised for failing to demonstrate the practical applicability and usefulness of their ideas to non-academics. As A. Johnston of the Economic Section commented:

I quite see that it is for economists to say 'here are the general principles' and that it is for the various boards to say how far they can apply them but I feel that we ought to make a greater endeavour than has been done hitherto to indicate how any general principles can be applied in practice. We have to face tremendous mental inertia in the practical men and a great level of thinking on their part derives from the time when they were engaged in exploiting a monopoly.[44]

Nor was there always much tolerance shown towards the more theoretical economists within government by senior civil servants who were trying to get through the immediate crises and problems. Confiding to his diary, Robert Hall described one eminent economist working within government as:

a little lazy and has fits of academic conscience in which he feels that the truth as he sees it is more important than anything else. This would be all right if he took the trouble to check his facts and his reasoning but like all economists he is an escapist and thinks he is being virtuous in taking a high line when he really hasn't done enough to justify any line.[45]

Yet, even when other economists did address the practical problems of resource allocation in a high, and often excess, demand economy, the level of understanding and sympathy with which their ideas were treated by politicians, including ministers, was less than encourging. The recommendations of Professor Arthur Lewis and his three supporters on the Ridley Committee,[46] for an excise duty of £1.00 per ton on coal, were ignored. Even when an economically literate minister like Gaitskell advanced proposals for differential coal prices which had already been approved by the Cripps-dominated Production Committee, these proposals were talked out in the Economic Policy Committee, with James Callaghan (Parliamentary Secretary, Ministry of Transport) arguing that not only would such increases add a further £1.5 million to the railway's coal bill which had already risen from £12.5m in 1938 to £37.5m in 1948, but that also any consequent increase in railway freight charges would simply bring discredit on the government's policy of nationalisation.[47] When the use of the price mechanism to choke off and separate out sections of demand met with such a response, it was hardly likely that arguments for sectoral co-ordination based on cost-pricing policies would be welcomed. Indeed, they were rarely understood. When Meade sought an interview to discuss Chantler's proposals for introducing cost-based pricing as the basis for the co-ordination of the transport sector, Morrison was outraged, accusing Chantler of wanting 'competition, not co-ordination'. In the end, Meade had more success arguing from the industrial efficiency angle, saying that it was a waste of labour if goods were taken on a system with a higher marginal cost, since this reduced output per head just as did technical inefficiency in any industry. This began to make some impression.[48]

The problems experienced in winning acceptance of the need to make greater use of price mechanisms in allocating resources had a series of effects on the operation and performance of the nationalised industries. Most industries continued to operate in conditions of excess demand but without being able to accumulate financial reserves. In turn, this made it difficult to finance capital investment programmes (which in turn were inflated in size by the excess demand), a problem exacerbated by the fact that book costs tended to be depreciated on a historic rather than a replacement cost basis. In the absence of any basic relationship being established between resource costs and prices, any attempts at co-ordination within sectors were likely to become increasingly administrative. Quite how administrative allocations were to be made without some basic information on opportunity costs was difficult to see. Indeed, the general ignoring of price mechanisms both accompanied and encouraged the ignoring of the benefits of a decentralised cost and price structure in which prices in industries like coal would reflect variations in costs between regions and products. The tendency implicit in ignoring the scope for a fuller use of the price mechanism was to strengthen further the tendency to centralise in the hands of administrators, decisions which might have been left at a local level. As James Meade noted:

> In the absence of a properly functioning price system to serve as a guide, the managers of plant in socialised industries would lack criteria for the economic conduct of their enterprises and it would become necessary to transfer to the centre the responsibility for many types of decisions which could otherwise, with advantage, have been taken locally.[49]

The centralising tendencies contained in the statutory legislation and in the rejection of a greater use of market mechanisms within the nationalised industries, made the external supervision and appraisal of each nationalised board's decisions of especial importance. Suggestions that independent committees be established to monitor and examine central decisions on such aspects of management as tariff structures were not treated seriously, such independent committees being viewed as detracting from the independence and authority of the national boards themselves.[50] The main supervisory role therefore fell to the sponsoring department, but these departments were often ambivalent in their attitude to such a function and limited in their ability to exercise it fully. Not wishing to duplicate the functions and staff of the nationalised industry, the scope for government intervention was limited as much by its lack of information and expertise as by its wish to preserve 'management's right to manage'. Not only did Gaitskell recognise that any regular interference would reduce the morale and standing of the boards, and 'deter industrialists, who detest the thought that they will be becoming civil servants doing ministers' will'[51] but he also saw that asymmetries in the distribution of information and expertise between government and the nationalised industries left government ill-equipped to advise industries on the more technical aspects of their decision-making:

> We should not ... attempt to say which mines should be developed and which should be closed, what form of generating station ought to be constructed, nor what system of gas distribution would be appropriate in a particular area — though these, be it noted, are all issues of major policy for the Boards themselves.[52]

This may well have been a realistic recognition of the impracticality of government intervention on such technical matters.[53] It did, however, mean that if many of the benefits for fixed capital investment which had been associated with nationalisation were to be realised, then, first, these boards should be seeking to pursue fixed capital investment programmes of the highest efficiency, and, secondly, where they did not, that the government would have both the will and the ability to bring them back into line. Thus, recruiting managers and board members who were both able and sympathetic to the general aims of nationalisation became particularly important and particularly problematic.

That the government should have experienced problems in finding enough managers capable of operating a nationwide enterprise is not particularly surprising. Even in the inter-war period, finding managers to run large corporate enterprises proved difficult, recourse often being made to the railways, government (in particular the Inland Revenue) and the armed forces for personnel who had experience of running such large organisations. In the post-war period of excess demand, increasing industrial concentration and relatively low unemployment, recruiting talented managers was almost certain to be difficult. As Gaitskell noted in a Labour Party memorandum on the administration of the nationalised industries:

> Unfortunately, it is exceedingly difficult to get men of the right brains, experience, personality and knowledge of men to fill the top jobs either on or just below the Boards. There is a sort of idea among some of our friends that 'one man is as good as another'. No idea could be more remote from the truth. Upon finding the right man for these jobs depends, more than on anything else, the success or failure of these experiments. . . .[54]

In the competition to recruit skilled managers, one obvious strategy was to offer attractive wages. Yet, in the case of nationalised industries, the scope for any such strategy was constrained by political/public relations considerations. When Shinwell proposed paying salaries of £10,000 to the NCB Chairman, £7.5–8,000 to the Deputy Chairman and £6,000 each to the board members, in order to attract talented men, the Lord President condemned these salaries as being 'socially objectionable', while the Chancellor of the Exchequer agreed that paying a five figure salary to the NCB Chairman would not make for good public relations. Instead, it was suggested that the Chairman should be paid £8,500, the Deputy/Vice Chairman £7,000 or £7,500 if absolutely necessary, and other members £5,000 each, but that expenses might be on a liberal scale.[55]

Once appointed, the government was keen that such nationalised boards should continue to act in the 'national interest'. Quite what this meant was unclear, but what did become clear was that a board's and the government's perception of what constituted the 'national interest' could vary significantly. Most familiarly, the electricity industry saw itself as acting in the national/public interest when it pursued an aggressive expansionary programme designed to bring cheap electricity to as many people as possible. That this might conflict with government efforts to reduce demand and that its fierce competition with the gas industry might conflict with notions of co-ordination does not appear to have weighed heavily on the industry's conscience. Unwilling and often unable to make greater use of pricing mechanisms and criteria, and confronted by a monopoly holding a predominant share of information and expertise, the government's ability to influence a nationalised industry was limited.[56] It was also politically difficult for the nationalising Labour Government to appear to be involved in open hostilities with one of its nationalised creations.[57] Personal contacts, lunches, face-to-face meetings and committee discussions between Government and nationalised industry managers were obvious avenues of influence, but their scope was limited and the participants were sometimes dissatisfied. Ministers and officials often complained that managers simply reiterated well-known arguments, while board Chairman frequently complained that they were not properly consulted.

That the bureaucratic mechanisms for information-exchange and decision-making were producing problems was worrying, particularly given the onus which the rejection of a greater use of market mechanisms placed on bureaucratic processes of decision-making. The limited use of market mechanisms and the apparent inadequacies of bureaucratic control were of particular concern, given the size and intensity of the nationalised industries' use of capital investment resources. With price controls inflating demand and cross-subsidisation obscuring costs, the potential for sinking ever more resources into demand-led investment programmes was clear. 'Crowding-out' was as much, if not more, of a threat on the demand-side as on the financial supply-side. Moreover, unless effective bureaucratic means could be devised for checking the expansionary programmes of industries like electricity, then the ability to realise the early hopes for co-ordination within sectors was unlikely to be realised. This lack of co-ordination was already of concern by 1951. As the TUC noted in its evidence to the Ridley Committee:

> The lack of co-ordination between the development plans of the fuel and power industries must give rise to serious concern over the continued absence of comprehensive policy for the industries as a group ... [the] fact that these industries have followed their separate paths emphasizes that the opportunity under public ownership of adequate co-ordination will diminish as these projects get under way.[58]

The TUC did acknowledge 'the difficulties of the transitional stage to public ownership and the first year or two afterwards', but also urged that the time for tackling this problem is now: 'further delay can only intensify the difficulties which have already arisen'.[59] Yet although economists like Meade and politicians like Gaitskell regarded the initial organisational structure of the nationalised industries as 'transitional' and 'experimental', once established the structure proved very difficult to alter. Indeed the asymmetries in the distribution of information and expertise between nationalised industries and government which derived from the centralised organisational structure of monopoly industries contributed significantly to allowing the industries to fend off attempts at radical structural reform. Thus, the early post-war period saw the establishment of many of the guidelines and practices which were to shape the nationalised industries in years to come. The decision to opt for monopolies in major industries was beginning to be questioned within the Labour party by the early 1950s but no radical departure from monopoly thinking occurred for several decades after that. The resistance to making greater use of market mechanisms within nationalised industries, a resistance which stemmed from inter-war perceptions of the operation of markets as well as the sheer abstraction and at times irrelevance of the post-war average/marginal cost pricing debates among economists continued. Prices continued to be distorted and sectoral co-ordination became an administrative nightmare. Governments continued to experience the serious effects of asymmetries in the distribution of information and expertise and, in an attempt to improve its control over the nationalised industries, increasing use was made of external financing limits. Such control could only ever be partial. Many of the problems which the nationalised industries encountered in the forty years after nationalisation can be traced back to the decisions and debates concerning 'competitiveness' and competition in the 1930s and 1940s. Thinking and practices established then proved very difficult to change.

Notes

1 Sir Alec Cairncross, *Years of Recovery: British Economic Policy, 1945–51*, London, 1985, p. 476.

2 When discussing the draft bill for the nationalisation of the coal industry, the Minister of Supply did consider prohibiting the importing of coal by any one other than the National Coal Board. See Sir Norman Chester. *The Nationalisation of British Industry, 1945–51*, London, 1975, p. 189.

3 Barry Supple, *The History of the Coal Industry, Volume 4, 1913–46: The Political Economy of Decline*, Oxford, 1987.

4 T. R. Gourvish, *British Railways, 1948–73: A Business History*, Cambridge, 1986.

5 Labour Party Post-War Finance Sub-Committee, RDR 267/April 1944, 'Full employment and financial policy'.

6 Ibid.

7 For an account of the work of the Economic Section, see Alec Cairncross and Nita Watts, *The Economic Section 1939–61: A Study in Economic Advising*, London, 1989.

8 'The socialisation of industries. Memorandum by the Economic Section of the Cabinet Secretariat'. See Susan Howson (ed.), *The Collected Papers of James Meade, Volume 2, Value, Distribution and Growth*, London, 1988, p. 75.

9 Labour Party memorandum, RD33/November 1946, 'Criteria for Nationalisation'.

10 *Ibid.*

11 Other considerations included safety and the public interest. See *Ibid.*, 'In some cases, it may be impossible to safeguard the public interest adequately without public ownership. Thus, it may be too dangerous, as with atomic energy or water (where purity is important to health), to allow private capital to exploit a particular service; or at least the necessary inspectorate to supervise private industry may sometimes be far too expensive and cumbersome to tolerate.'

12 A. Beacham, 'Efficiency and organisation of the British coal industry', *Economic Journal*, LV, 1945, p. 214.

13 See, for example, the discussion of the Brassert consultancy reports on the inter-war steel industry in Steven Tolliday, *Business, Banking and Politics: The Case of British Steel, 1918–1939*, Harvard, 1987.

14 Labour Party memorandum, 'Criteria for nationalisation'.

15 Labour Party memorandum, RD 295/June 1949, 'Development of nationalised industries: copy of letter and memorandum from the Lord President to the Minister of Fuel and Power, 17 March 1949.'

16 See Chester, *Nationalisation*, pp. 191–2.

17 See G. B. Richardson, *Information and Investment*, Oxford, 1960. Also see the discussion of Keynes' thinking on the relationship between capital investment and uncertainty in D. E. Moggridge, *Keynes*, London, 1976, chapter 5.

18 Labour Party memorandum, RD140/August 1948, 'British steel at Britain's service'.

19 Labour Party memorandum, RD1/September 1945, 'A Labour policy for privately owned industry'.

20 Labour Party memorandum, 'Criteria for Nationalisation'.

21 Tom Wilson, for one, thought that the presence and potential of economies of scale was greatly overrated. See Thomas Wilson, *Modern Capitalism and Economic Progress*, London, 1950, pp. 155–6.

22 *Reid Committee Report on the Coal Industry* (Cmd. 6610, 1945), *McGowan Committee Report on Electricity Distribution* (1936–7), *Heyworth Report on the Gas Industry* (Cmd. 6699, 1945).

23 Cairncross, *Years of Recovery*, pp. 468–9.

24 Chester, *Nationalisation*, p. 1027.

25 *Ibid.*, p. 1033.

26 *Ibid.*, p. 434.

27 *Ibid.*, p. 419–20.

28 *Ibid.*, p. 1032.

29 *Ibid.*, p. 434.

30 Meade, *Collected Papers*, Volume 2, p. 53.

31 *Ibid.*, p. 64.

32 Chester, *Nationalisation* p. 414. The Ministry wanted autonomous, highly independent boards for distribution and Morrison wanted a single national authority for generation and distribution.

33 Chester, *Nationalisation*, p. 418.

34 See 'The management of socialised industries', in Herbert Morrison, *Socialisation and Transport*, London, 1933, pp. 131–48.

35 Among other arguments used by Morrison in pushing for centralised structures was the general difficulty of finding sufficient managers and technical staff capable of manning a structure in which decision-making was dispersed and decentralised.

36 Labour Party, RD254/January 1949. Memorandum on the 'Administration of nationalised industries: comment by Hugh Gaitskell on draft report of subcommittee', circulated to the Labour Party Sub-Committee for Information. Also see similar comments by James Meade, *Collected Papers Volume* 2, p. 61.

37 Labour Party memorandum 'Administration of nationalised industries'.

38 T. Wilson, 'Price and output policy of state enterprise'. *Economic Journal*, December 1945, pp. 454–61. 'If you are happy to take losses, how do you ensure efficiency?' Wilson notes that Meade is prepared to do without a balance sheet for 'the amount of profit or loss made is irrelevant'. He suggests, it is true, that it may be possible to draw up production functions for different plants which would make it possible to spot inefficient managers who could then be dismissed. As Mr Fleming rightly notes, 'this is likely to remain a pious hope'.

39 Meade, *Collected Papers*, Volume 2. p. 62. 'The basic function of prices is to guide production towards those goods and services which are most needed by the community, and to ensure that production is carried out by the most economical methods'.

40 Meade and Lerner (as described by Wilson) felt that even when losses were being made, marginal costs and prices should still continue to be equated, the losses being met out of government subsidies. As Wilson argued, this marginal policy was also likely to extend the number of industries coming under socialisation, since 'it is rightly regarded as inconceivable that the State should subsidise a private monopoly in the hope that it would then dutifully keep the commandments.' Given a wide application of marginal cost pricing rules, and the fact that increasing returns are likely to be fairly widespread, it seemed logical to expect the extension of a large programme of socialisation beyond the public utility field. T. Wilson, 'Price and output policy of state enterprise'.

41 R. H. Coase, 'Price and output policy of state enterprise: a comment', *Economic Journal*, April 1945, pp. 112–13. The illustration given by Coase was that where an electricity generating plant was installed for the supply of one country house. Following Meade's 'Rule', the occupier would only be charged on the basis of marginal cost, the remainder of the cost being borne by the rest of the community. However, while not disputing the fact of any such inequality, its extent would be less than Coase suggests, since the new generating set would be likely to be run at full capacity as a basic contributor to meeting base-load requirements. At peak-hours, it would be the older sets that would be brought on-stream.

42 *Report of the (Clow) Committee to Study the Peak-Load Problem in Relation to Non-Industrial Consumers*, Cmd. 7464, July 1948.

43 Meade, *Collected Papers, Volume 2*, p. 62.

44 PRO T230/28, Economic Section of Cabinet Secretariat, *Discussion Papers 1948*. Paper by A. Johnston (Economic Section) on the 'Problems of socialisation', 8 June 1948.

45 Alec Cairncross (ed.), *The Robert Hall Dairies, 1947–1953*, London, 1989, 16 April 1948.

46 *Report of the Committee on National Policy for the Use of Fuel and Power Resources*, September 1952, Cmd. 8647.

47 PRO CAB 134/216, Economic Policy Committee. EPC(48) 28th meeting, 9 July 1948.

48 Meade, *Diaries*, British Library of Political and Economic Science, 17 March 1946.

49 Meade, *Collected Papers, Volume 2*, p. 64.

50 *Ridley Report*, paras 62–3. Also see D. L. Munby, 'The price of fuel', *Oxford Economic Papers*, 1954, N.S. 6, p. 242, footnote 4.

51 PRO CAB134/690, Committee on Socialisation of Industries, SI(M)(49)33, 30 May 1949, 'Government control over socialised industry', memorandum by Minister of Fuel and Power.

52 *Ibid.*

53 In May 1949, Gaitskell instanced in a memorandum to the Committee on the Socialisation of Industry occasions on which he had intervened. Recently he had intervened to prevent the NCB making an agreement with the merchants to establish fixed wholesale margins in the trade. Both NCB and merchants wanted such margins and were prepared with a plan which would involve the Board depriving a wholesale merchant of supplies who sold to a merchant at less than the agreed fixed margin. Gaitskell also intervened to persuade the NCB against their wishes to increase exports of anthracite to Canada, although this involved a price 25s below what they could have obtained in European markets. The Board, persuaded of the overriding importance of exports to Canada agreed.

54 Labour Party memorandum 'Administration of nationalised industries'.

55 PRO CAB 134/687 SI(M)(46) 2nd meeting, 15 February 1946. The issue was still rumbling in 1947. By then Dalton was generally obtaining Attlee's agreement that Chairmen of major boards should be paid £7,500–£8,500 p.a. with full-time board members receiving £5,000 (or a little less) per annum. However, Attlee still continued to be concerned at the long-run impact which such salaries would have, not so much on industry as on the civil service. For example, the maximum that could be paid to a Divisional manager of the NCB was greater than that paid to the Permanent Secretary of the Treasury, and only £1,000 less than that of five Cabinet ministers.

56 The problem of the 'Public Control of Monopoly' came increasingly to concern the Labour Party. Labour Party, RD 44/February 1947, 'The public control of monopoly'.

57 Gaitskell hoped that the influence of this political factor would decline over time. See SI(M)(49)33, 30 May 1949, 'Government control over socialised industry', memo by Minister of Fuel and Power, Hugh Gaitskell.

58 Ridley, Appendix XII(b).

59 Ridley, Appendix XII(b).

5 *Helen Mercer*

The Monopolies and Restrictive Practices Commission, 1949–56: a study in regulatory failure

The Commission will be dealing with one of the toughest industrial problems of our time and investigating some of the most powerful monopolies and combines in the country.[1]

1 Introduction

This chapter looks at the attempt to use a regulatory body, the Monopolies and Restrictive Practices Commission (MRPC), from 1949 to 1956 to remove deficiencies in Britain's industrial structure and so to raise British competitiveness.

The MRPC was established under the 1948 Monopolies and Restrictive Practices (Inquiry and Control) Act. This in turn was the enactment of a promise in the Employment Policy White Paper of 1944, which noted that:

There has in recent years been a growing tendency towards combines and towards agreements, both national and international, by which manufacturers have sought to control prices and output, to divide markets and to fix conditions of sale.[2]

The White Paper went on to argue that, 'Such agreements or combines do not necessarily operate against the public interest; but the power to do so is there.' The paper therefore promised to investigate restrictive practices and the activities of combines to check practices which 'may bring advantages to sectional producing interests but work to the detriment of the country as a whole'.

The Act of 1948 established a system of enquiry on a case by case approach. If the Board of Trade had reason to believe that monopoly existed it could refer the case to the MRPC. Monopoly was defined as a situation where at least one-third of the goods was supplied by one firm or by two firms acting together in such a way that competition was restricted. The Commission then reported either simply on the facts of monopoly conditions in the industry or product or it could assess the relation of the practices to the 'public interest', which was only very loosely defined. Should the report be adverse the Board of Trade, or another responsible Ministry, could take remedial action, which at most was the declaration by Order in Parliament that the practices were illegal.

The chapter argues that this experiment was a failure, because the very organisations which the Commission was charged to control — trade associations and large firms — exercised a decisive veto on both the nature of the legislation establishing the Commission and on the administration of the work of the Commission by the Board of Trade. The chapter concentrates especially on the administration of the Monopolies Commission and the relations between government and industry illustrated in the choice of references to the Commission. It concludes that the Commission provides a British example of regulatory failure: industrialists nearly immobilised the administration of regulation.

It is clear that these conclusions have important implications. Current government strategy stresses the role of an open, competitive economy in improving British performance.[3] Moreover, as regulatory bodies for privatised sectors replace the nationalised industries, fears of regulatory failure by these agencies are widespread.[4] Yet Labour's alternative strategy places great reliance on the power of regulation.[5] This chapter therefore comments on the historical experience of regulation in Britain.

2 The problem of British restrictive practices

Monopoly and cartelisation had come to dominate some of the largest and most important sectors of British industry. The largest 100 firms accounted for perhaps 23 per cent of manufacturing output in 1935, and 26 per cent by 1953.[6] It is well known that the rate of increase of concentration slowed in the 1930s and 1940s, and it has been assumed that alternative paths to market control were taken through the use of cartels.[7]

It is, of course, very difficult to estimate the extent of cartel activity. British cartels were, quite properly, called restrictive practices, in part because they lacked, at least until the 1950s, the formalisation of the typical German variety. The British 'cartel system' in the 1930s was characterised, apart from exceptions in coal, cotton, iron and steel, and shipbuilding, by a mass of petty, defensive restrictions, gradually becoming more organised and formal during the war and post-war periods.[8]

Quantitative assessments made in recent years indicate that between 25 and 30 per cent of gross manufacturing output in the 1930s was controlled by trade association cartel arrangements. Analysis of the register of restrictive agreements established in 1956 indicates that this figure had risen to between 50 and 60 per cent of manufacturing output in the mid-1950s.[9] It is difficult to distinguish between an increase in available evidence or a real increase in the extent of cartelisation. Some consensus does exist, however, that cartelisation strengthened between the 1930s and the 1940s to 1950s.[10] This was as much a result of the impact of tariffs in the 1930s, and policy during and after the war, using trade associations to adminster wartime controls of industry, and as agencies representing industry on, for instance, the Central Price Regulation Committee.

Restrictive practices flourished across all sectors of manufacturing industry. In the 1930s areas particularly affected were in iron and steel production and manufactures, non-ferrous metals and building materials. By the 1950s these were joined by products in electrical engineering, textiles and food, drink and tobacco. The types of practices most common by the 1940s and 1950s were simple price-fixing or common prices, resale price maintenance and the use of exclusive dealing, collective boycott, stop lists and private courts, and uniform or level tendering, especially for contracts for local government, nationalised industries and the health service. This last practice had become something of a public scandal by the 1950s.[11] The redundancy agreements of the cotton, shipbuilding and other industries in the 1930s were less frequent, although the Commission investigated a number of agreements to restrict capacity. Britain was still participating in a number of important international cartels well into the post-war period.

Policy to deal with restrictionism had been high on the agenda of wartime reconstruction discussions. For economists the inherently restrictionist nature of British cartels in the 1930s was a main obstacle to an expansionist post-war economy and hence to plans for full employment. In line with modern thought cartels were seen as more insidious than 'combines' in that they exploited monopoly positions, raised profits and prices and preserved the inefficient without having any possible compensations in the form of economies of scale.[12]

The strongest official condemnation of cartels came in the report of the post-war export trade committee, which blamed extensive price rings for Britain's poor competitiveness in unprotected markets. Price rings were accused of basing prices on the less efficient producers, of preventing the full use of modern factories, and of raising the prices of components used by export industries, for instance in the car industry, requiring iron and steel, electric bulbs, glass and alloy steels.[13] Thus concern for British competitiveness was an important feature of wartime discussions, especially as it was linked with questions of full employment and an expansionist economy.

Efficiency was seen as a strong basis for improved competitiveness. The one clear policy for private industry in the Labour Party's 1945 manifesto promised public supervision of monopolies and cartels 'with the aim of advancing industrial efficiency in the service of the nation'.[14] In introducing the bill into parliament, Harold Wilson said that the nation could not afford restrictive practices if they increased the cost of British products abroad, prevented the development of inventions and new techniques, reduced output, maintained excessively high prices or denied entrance to an industry by an efficient competitor.[15] Many Labour speakers in the debate supported the idea that control of restrictive practices would 'remove abuses and restore the competitive power of British industry'.[16] Speakers on both sides had some measure of agreement that practices which restricted production, when the priority was to expand it, had no place.

There was a clearly-stated concern with British *competitiveness* in world markets, but this was not synonymous with the encouragement of *competition* within British industry. Much of the Labour government's actions, for instance in the Development Councils and nationalised industries, was to encourage amalgamation, rationalisation and co-operation. Some economists may have seen the 1948 Act as an expression of a growing commitment to the virtues of competition,[17] but it was not until 1956 that the President of the Board of Trade, Peter Thorneycroft, introduced the Restrictive Trade Practices Bill with an ideological statement on 'the virtues of free enterprise — initiative, adaptability and risk-taking',[18] into which the cause of industrial efficiency was implicitly subsumed. For most of the period of operation of the MRPC the ostensible problem being addressed by the government was restrictive practices as a block to industrial efficiency and competitiveness. Competition was one possible means to this end: as Dalton put it: 'you should either have really free, profit-seeking competition, or else a centrally-planned public enterprise'.[19]

However, the Coalition and Labour governments' commitment to competitiveness weakened in the face of industrialists' opposition to moves against their practices. The mass of articulate businessmen publicly argued that the cartels of the inter-war period were a way of improving Britain's export position. International cartels had preserved markets which might otherwise have been lost, and domestic cartels were a rational and orderly way of conducting business.[20] Plans made during the war to abolish certain practices, or to establish a procedure of registration and enquiry, not dissimilar to that of the 1956 Act, were dropped for fear of antagonising industry. Instead the Labour Government, guided by their officials, opted for an approach whereby industrialists were to be gently shown that their arrangements did neither themselves nor the country any good. As Wilson expressed it in the Second Reading, he hoped that the publicity engendered by a report 'will cause any of those industrialists who may, without realising it, be acting in an anti-social manner to alter their arrangements'.[21]

Indeed, so strong was business opposition to the proposals that the Act might not have been introduced had it not been necessary to bow to American pressure for British co-operation to curb international and domestic cartels.[22] In addition, the government needed to assuage public opinion. As in the First World War, strong public suspicion of monopoly, trusts, cartels and profiteering arose.[23] By 1948 the nationalisation programme had dampened much of this feeling which so fuelled early American anti-trust legislation. But accusations of profiteering remained,[24] and the high profits of the immediate post-war years were later to be blamed on trade association cartels.[25] The government clearly intended to use public opinion as another weapon against industrialists, seeing publicity as the key sanction to make the Commission effective.[26]

Thus the aims, both real and concealed, may be summarised. An attack on restrictive practices and monopolies was to be mounted in the interests of industrial efficiency. But in the interests of peaceful government–industry relations, the attack was to be gradual. Stringent measures like price controls, outlawing certain practices, or imposing public supervision and ownership in cases of 'market failure' were rejected in favour of gentler forms of persuasion. In particular the government hoped that the potentially adverse publicity of each report would influence the behaviour, not only of the industry under enquiry, but of all industrialists. Yet even with the mild approach of the 1948 Act some administrators were worried lest the bill might 'rock the boat' of government–industry relations.[27]

3 The work of the monopolies and restrictive practices commission

We now turn to the administration of the work of the Commission itself, assessing its performance in the light of the initial objectives of regulating private industry in the interests of efficiency and competitiveness.

Regulation of the capitalist enterprise is little explored in British economic history. To date analyses of early agencies, for instance of the Import Duties Advisory Committee, or the first Monopolies Commission, have been limited and essentially geared to prescriptions for reform, either from an economist's or a lawyer's standpoint, or have been written by members of the regulatory agency as little more than self-justification or defences of government policy.[28] The approach of the majority of writers on British competition policy is 'Whiggish', viewing the development of policy as the result of the gradual awakening and improvement of popular, governmental and business appreciation of the problems and dangers of restrictionism.

Nowhere better is this approach indicated than in conclusions on the work of the first Commission. G. C. Allen, for instance, who wrote extensively on the monopoly problem in Britain and was an early member of the Commission, argued that:

> the proceedings and the reports of the Commission exerted a powerful influence on the attitude to restrictive practices both in board rooms and in the market place … Politicians, journalists and the general public became more closely acquainted with the activities of cartels and monopolies and with the informed criticisms directed at them. The former complacency was disturbed. The main achievement, therefore, was the creation of opinion critical of restrictive practices.[29]

American historians also at first viewed the development of anti-trust legislation as a process of 'reform' to guard the 'public interest' against the business community.[30] Since then, though, a substantial literature in the United States has revised this idea drastically. A body of opinion has come to predominate which, in various ways, and from opposite political corners, attacked regulatory commissions for becoming, or for always being, the captive of the very interests they were supposed to be regulating.[31]

Much of this literature focuses on the ability of business interests to modify the form of their regulation. As Marvin Bernstein says,

> The hostile environment in which regulation operates is pervasive. It provides the frame of reference for the regulatory process conditional upon the acceptance of regulation by the affected groups. It forces a commission to come to terms with the regulated groups as a condition of its survival.[32]

The first Monopolies Commission does provide an example of regulatory failure, but this is due less to the nature and behaviour of the Commission and its members, and more to the relationship between government and industry and the government's unwillingness to adopt a strong line which might antagonise industry. This attitude was evident from the outset, as noted above, in the legislation itself. The rest of the chapter looks at how it governed the work of the Commission.

The economic impact of the MRPC was minimal. To begin with there were very few referrals, and those that were made took an average of two years and three months to complete (see Table 5.1). One report took four years and one month and it was another eight months before it was published. Much of this slowness may be attributed to the difficulties of obtaining information from the firms involved.[33]

Even fewer reports were acted on by the government. This was partly due to the time it took to negotiate with the firms over how the Commission's recommendations were to be implemented. In addition, the forthcoming change in 1956, which made many agreements registerable and illegal, meant that the Board of Trade preferred to let the new Registrar deal with any practices which the Commission had criticised rather than court controversy with the firms involved.

The slowness of the Commission meant that the government had taken action over only nine of the reports where action was required, before work was halted with the introduction of the 1956 Act. In three cases there were still complaints, mentioned in the Board's annual reports, of the same practices continuing, or of the industry continuing to charge allegedly high prices.[34] In a fourth case, that of calico, the redundancy scheme, although criticised by the Commission, was allowed to run its course until 1959.[35]

Frequently the government did not implement Commission recommendations, for instance in the calico redundancy schemes or the aggregated rebates scheme in electric lamps, or price supervision of the British Oxygen Company or the British Match Company.[36] Sometimes economic conditions in the industry nullified the effects of any abrogation. Strong information

Table 5.1 *Summary of the work of the Monopolies and Restrictive Practices Commission, 1949–56*

Product investigated (supply of)	*Date referred month and year*	*Date report signed by MRPC*	*Report published*	*Action taken on MRPC recommendations*
Dental goods and Instruments	March 1949	Nov. 1950	Dec. 1950	Order made that practices illegal, July 1951
Cast-iron rainwater goods	March 1949	Feb. 1951	April 1951	Alternative arrangements from Jan. 1952
Electrical filament lamps	March 1949	Aug. 1951	Nov. 1951	Voluntary undertakings. Many later complaints
Insulated electric wires and cables	March 1949	April 1952	July 1952	Minimum prices allowed Jan. 1955 in line with MC conclusions
Matches and Match-making machinery, and exports of	March 1949 March 1949	Oct. 1952 Oct. 1952	May 1953 May 1953	New international arrangements made in March 1954
Copper and copper-based alloys, semi-manufactures and exports. FACTS ONLY	Dec. 1950	April 1953		Converted to public interest
Copper semis: Public Interest	July 1953	Jan. 1955	Sept. 1955	NONE
Insulin	Dec. 1950	July 1952	Oct. 1952	Not against public interest later investigation
Printing of woven fabrics (calico)	April 1951	Nov. 1953	April 1954	End some practices; redundancy scheme to run its course
Imported timber	Oct. 1951	July 1953	Oct. 1953	Fed. agreed to recommendations but later found not to comply
Electrical and allied machinery and plant and exports	April 1952	July 1956	Feb. 1957	NONE
Tyres, and exports	Sept. 1952	June 1955	Dec. 1955	NONE
New buildings in greater London, (costing more than £1,000 each)	March 1953	June 1954	Sept 1954	Practices ended 1954 and local authorities elsewhere advised to be aware of the problems
Hard fibre cordage	July 1953	Dec. 1955	Dec. 1955	NONE
Lino	Sept. 1953	Jan. 1956	Jan. 1956	NONE
Gravel and sand in central Scotland	Dec. 1953	Nov. 1955	March 1956	NONE
Industrial and medical gases	Feb. 1954	June 1956	Dec. 1956	March 1956: voluntary undertakings: No action to regulate price and profits
Standard metal windows and doors	Feb. 1954	Sept. 1956	Dec. 1956	NONE
Rubber footwear	April 1954	May 1956	June 1956	NONE
Electronic valves and cathode ray tubes	Dec. 1954 Aug. 1956	Sept. 1956	Dec. 1956	Report limited to facts –
Electric street-lighting equipment	June 1955	ENDED MARCH 1956		
Steel frames for buildings	June 1955	ENDED MARCH 1956		
Tea	July 1955	Sept. 1956	Dec. 1956	Practices not found against the public interest
Electric batteries	Sept. 1955	ENDED MARCH 1956		
Chemical fertilisers	Oct. 1955	July 1959	July 1959	Undertakings given
Section 15 references:				
Collective Discrimination	Dec. 1952	May 1955	June 1955	1956 legislation
Common minimum prices and level tendering	Oct. 1955	ENDED MARCH 1956		

Note: Table summarises reports on industries referred to MRPC before the 1956 Restrictive Trade Practices Act came into operation

Source: Monopolies and Restrictive Practices Acts, 1948 and 1953, Annual Report by the Board of Trade for years ending 1955, 1956, 1957, 1958, 1959

agreements replaced the formal price lists in calico, while price leadership by the dominant firm, BICC, eventually developed in electric wires and cables, although at lower price levels.[37] The Board of Trade used its powers to make an order outlawing the practices criticised only in the case of dental goods. Criticism of this action was strong, and the power was not used again in the lifetime of the old Commission, except as a threat to get industries to comply with the Commission's recommendations.[38]

Nor did the reports which were published have any major impact on the rest of industry. A paper presented to an interdepartmental group chaired by Sir Frank Lee to review the policy, noted how speakers from all parties in the debate in the 1948 Act had expressed a belief that responsible people in industry would use the Commission's findings as a guide to their own conduct. This, the paper stated, had not happened. Practices found objectionable in the supply of rainwater goods were common to the supply of many building materials. Arrangements similar to those in dental goods probably existed in the supply of scientific instruments and laboratory ware. Yet, the paper commented, neither the Ministry of Works nor the Ministry of Health had had talks with other trade associations in the light of the two reports. The reasons for the inertia were, the paper suggested, overwork and lack of experience so that 'most Departments will regard it as premature to court more trouble by broaching restrictive practice problems with their other industries'.[39]

Some responses to the work of the Commission have been noted.[40] But Pilkington, which did respond to the threat of action under British and Canadian legislation by reviewing its practices,[41] was later told by the Board of Trade that it was one of the few firms which had adjusted its activities to 'present trends of thought'.[42]

Overall this balance sheet points strongly to 'regulatory failure'. The only benefit most commentators have seen in the experience of the Commission has been the gradual awakening of public perceptions of the problem and the use of reports, especially a general report on collective boycott, to prompt stronger action in the 1956 Restrictive Trade Practices Act.[43]

However, even this conclusion is open to question. The origins of the 1956 Act are confused. There is no doubt that a desire for stronger legislation on the part of Peter Thorneycroft, the Labour Party, and some newspapers was a cause. It was encouraged by Commission reports, especially that on Collective Discrimination.[44] However, according to Lord Kilmuir, then Attorney General, the sole motive for the 1956 Bill was that he and Peter Thorneycroft decided that current legislation and the Commission were unpopular with *industry*, and the use of the courts would be more acceptable to industrialists while allowing the government to distance itself from the process.[45] Certainly incessant business complaints prompted the Board of Trade to reconsider procedure under the 1948 Act again in 1954,[46] and to promise changes in a debate in February 1955.[47] Thus the changes towards a judicial procedure, in which industries 'accused' would have greater rights to present their case, they believed, and where judges might be expected to have less doctrinal commitment to competition, sprang originally from business complaints. Nor is it clear that businessmen were converted by the Commission's reports. Their arguments in favour of their practices had changed little from the 1930s and 1940s, a point noted sadly by Jewkes while commending the Commission for its robustly liberal attitude.[48]

To date the causes of this regulatory failure have been looked for in the workings of the Commission itself. The Commission was small, the procedure slow, the investigations often insufficient and the notion of the 'public interest' was not coherently pursued.[49] But members of the Commission were certainly not impressed by businessmen's claims for their practices.[50] Recommendations could be far-reaching: a minority called for nationalisation in the case of British Oxygen. According to some American literature, the praise which an industry heaped on a regulatory commission was an indication that it had been 'captured' by that industry.[51] On this crude criterion the Commission enjoyed some success. Business pressure groups greeted many reports with a chorus of disapproval, charging the Commission with doctrinaire bias.

Where industrialists directed their most effective lobby was during those parts of the procedure handled by the Board of Trade, at the point of contact between government departments and businessmen. There was a clear division of functions between the Commission and the Board of Trade. The Board had the duty to identify cases for referral, which it did after consultation with the 'sponsoring' department. The Commission then conducted the enquiry, wrote the report and made recommendations. The procedure then returned to the Board of Trade and the sponsoring departments to decide whether any section of the report should be omitted on grounds of national security or business confidentiality, and then to conduct negotiations with the firms concerned with a view to carrying out the Commission's recommendations.

Industry was generally unsuccessful in getting critical parts of reports excised. Requests for excisions were a major irritant within the Board of Trade, but industries' arguments were usually dismissed.[52] Although industry might object and although negotiations were protracted the government usually secured voluntary abrogation of the agreements or changes in the constitution of associations to remove objectionable features. For many of the trades where the industry was beginning to look obdurate, as in copper semimanufactures, tyres and electrical machinery, the government escaped a major battle as the new legislation made action less pressing.[53]

The most important failure occurred when decisions were made over which items to refer. The choice of references determined the overall nature of the problem being presented to the Commission, hence to industry and to the public. This weapon of publicity was to prove a very good sanction against industry, who clearly did not like to be under the spotlight. It was in fact so good that the government had to be very careful indeed how it was used, and was thus open to pressure.

Indeed so unwilling were government departments to court controversy that by 1953 it was clear that the whole procedure was becoming unworkable as possible references dried up. The problem had first been noted in 1952.[54] In 1953 the problem had actually become more acute with impending legislation to increase the size of the Commission and therefore the volume of its work.[55] By mid-1954, with the publication of some extremely critical reports, most notably those on cables and calico processing, something of a crisis had arisen which threatened to discredit the whole empirical approach and to prompt demands for sweeping general prohibitions.[56]

There was no doubt in the minds of the civil servants involved that suggestions for referrals from departments were limited out of fear of antagonising their industries. As A. C. Hill, the Principal at the Monopolies branch of the Board of Trade, put it:

> there has been evidence of a growing hostility in some commercial and industrial circles towards the Commission and indeed towards the whole process of enquiry and control provided for by the monopolies legislation. This critical attitude has manifested itself not only in some recent letters to the President but in, for example, the remarks of directors at Annual Meetings and in the 'tough' attitude shown by industries in their discussions with Monopolies branch on points of detail. It seems very likely that this critical atmosphere has influenced departments in their attitude towards the making of new references.[57]

This unwillingness to antagonise industry permeated to Cabinet level.[58]

Still more remarkable than the problems of getting references at all were the actual products chosen. Indeed the peculiarities of the choice of references has puzzled commentators:[59] why were calico printing, sand and gravel in central Scotland and tea chosen, and not the Yarn Spinners' Association, cement or oils and fats? A complete picture of the reasons for each referral has been limited by the nature of the records but it is possible to make some generalisations about why some industries escaped, and others were put into the firing line.[60] It is to this key limiting element in the procedures that the rest of this chapter now turns.

There were no agreed criteria for the choice of reference, and the publicly-stated criteria were different from those outlined in departmental memoranda. The former stated that items were chosen to give the Commission a range of types of practices and industries. The latter, however, advised production departments to select cases on this basis but also on such criteria as the level of public complaint, where there was reason to think efficiency could be improved or production expanded, or where there was evidence that agreements adversely affected prices and the volume of exports.[61] Thus when questioned by the Select Committee which enquired into the working of the Commission in 1953, A. C. Hill admitted that, 'I would find it awfully difficult in regard to any particular reference to weigh the various factors which led to the choice of reference'.[62]

The Board of Trade paid little attention to public concerns in practice. A much-vaunted complaints procedure was irrelevant in determining which items were chosen (see Table 5.2). What generally happened was that the *absence* of public complaint was sufficient reason against a referral, but its presence was irrelevant to the final decision. One case provided by the Co-operative Societies, electric lamps, was pursued, and public complaints carried more weight where local authorities were involved, tyres, structural steelwork and street-lighting. Indeed, by 1953 the Board of Trade was sensitive to complaints that the complaint procedure was being ignored and advised departments to suggest industries mentioned in the annual reports, 'if only to dispel any idea that we keep off potentially explosive subjects and prefer only those which are likely to yield reports of a fairly anodyne nature'.[63]

Table 5.2 *Complaints to the Board of Trade of monopoly or restrictive practices, 1949–56*

There were seven annual reports by the Board of Trade from 1949–56, on the work of the Monopolies Commission.
The following items were mentioned as subjects of complaint four or more times.

Item	*Number of complaints*	*Nature of restriction*	*If referred*
Newspapers and periodicals	7	a	
Motor vehicles	7	a	
Agricultural implements	7	a	
Chemists' goods	6	a	
Electrical accessories and appliances	6	a, c	
Plate glass	6	a, b, c	
Gramophone records	5	a, c	
Petrol	5	a, b, c, d	
Structural steelwork erection	5	b	Yes but ended
Tyres	4	a, b, d	Yes
Electrical components for motor vehicles	4	a, c	Yes
Textile-making machinery	4	c	
Wallpaper	4	a, c	
Carpets	4	a	
Typewriters	4	a	
Electric street-lighting equipment	4	b	Yes but ended

Key to restrictions
a: in distribution, e.g. discriminatory arrangements
b: minimum price-fixing, restrictions on new entrants, level tendering in manufacturing
c: products under monopolistic control through large firms
d: under monopolistic control through trade associations

Source: Annual reports on the Monopolies Commission.

A second feature which emerges is that on the whole large dominant firms escaped the spotlight. In part this position was achieved through the very terms of the 1948 Act and the Labour government's bias in favour of the large firm. Intrinsic to the 1948 Act was the absence of provisions for dealing with monopoly abuse exercised by the large, oligopolistic firm. Thus it could be 'embarrassing' for government policy to get an adverse report and to do nothing about it, and for it to be revealed that there were few sanctions against the large firm.[64] Also intrinsic to the Act was the presumption in favour of references involving price rings and cartels rather than large firms, and this was continued in directives to departments from the Monopolies branch.[65] For these reasons the objection that the item concerned was controlled by a single-firm monopoly was used against the referral of metal containers, gas mantles, razor blades, sewing thread, perspex and nylon yarn.

The large firm was partly protected by policy decisions that 'insignificant' items should not be preferred,[66] for it was sometimes precisely in limited fields of production that a large firm might exercise a monopoly, for example in cellulose film, chain cables, salt, glues, glycerin, rubber belting, rubber thread and lighter flints. Sometimes the presence of competition in an industry, even one dominated by a few large firms, was sufficient argument against referral, for instance in the cases of alcohol solvents, pesticides, photographic materials and rubber chemicals.[67]

But more important were the strategic positions in the economy which many dominant firms occupied. There was particular reluctance to refer cases involving an international cartel. In the case of electric cables the international ramifications were specifically excluded because, 'It was quite possible that if this agreement were examined it would become clear that there was a conflict between our national interests, interpreted strictly, and our international commitments'.[68] Concerted lobbying by the Foreign Office and the Ministry of Fuel and Power against the referral of any aspect of petrol distribution in Britain was successful,[69] although this was an item with much support for an enquiry.[70]

Other industries occupied strategic positions of a different nature. The government was relying on investigation and publicity as a way of educating business opinion, and was by extension relying on a sympathetic attitude towards the Monopolies Commission by the mainstream press. Yet the newspaper industry was itself subject to monopoly influences.[71] It was accordingly raised as a possible item on several occasions but killed by civil servants, on one occasion because 'there might be some advantage in not provoking unfavourable press comment about the Commission's activities until there has been more time to accustom the press to thinking and speaking favourably of the Commission'.[72]

Large firms might be important in providing items for the defence programme, like ICI, and the government did not wish to harm its relations with them.[73] Industries which were considered to be efficient and good export earners were also protected from enquiry, and this argument served again to shield larger firms, for instance in the cases of metal cans and containers, china clay, flat glass, soap, phosphates, perspex, industrial solvents, and a number of basic chemicals.

The position of some firms was so vital to the very structure of an industry, or of British industry in general that officials sought to defend them against the 'pillorying' by the Monopolies Commission, or to defend themselves against the industrial opposition which such firms could mount. The case of ICI, a firm with strong involvement with government, and official concern in its affairs from its inception,[74] is instructive. Few cases of chemicals were referred: only fertilisers, a market controlled by ICI and Fisons, and medical and industrial gases, a monopoly not of ICI's, but of the British Oxygen Company. Yet the possible number of chemical cases considered was legion, prompting one official to remark that the difficulty with ICI cases was knowing where to begin.[75] There were many reasons for not looking at ICI but one forceful argument, made against referring caustic soda, soda ash, glycol and other heavy chemicals was that, 'we do not want the Commission to cut its teeth on ICI.'[76] The damage to the government's relationship with ICI if alkalis were referred has already been mentioned and to refer oil baize, they argued 'would bring us up against ICI and British Plastics'.[77] ICI's monopolistic position in many areas was very sensitive as the firm regularly featured on Labour Party 'shopping lists' for nationalisation.

A further problem with referring large firms was that they could easily complain of being 'singled out' or 'made a guinea pig'.[78] Thus it was that when the Board decided to refer Unilever's monopoly of non-mineral oils and fats, they decided to refer at the same time the fertiliser aspect of ICI's operations, 'to avoid charges of discrimination against one particular giant'.[79] Thus, even where large firms were involved in restrictive practices they could frequently escape enquiry.

A third feature which emerges is that other government policies might run counter to competition policy. The most notable example of this is the case of iron and steel where a newly-privatised industry argued that the sale of further firms might be jeopardised were certain pricing arrangements brought within the scope of an enquiry into the general practice of common prices and uniform tendering.[80] The lobbying by the iron and steel board was one factor in the demise of that reference. In the case of the yarn spinners the long-term problems of the cotton industry persuaded civil servants that nothing should be done to make their problems any worse. Some Ministers hoped the cartel would 'crack' under emerging competitive pressure anyway.[81]

Fourthly, in looking for reasons why items *did* get referred the most important seem to be either pressure from industrial customers of the trade concerned or complaints from overseas.

In the case of fertilisers, the consumers desirous of an investigation were the National Farmers' Union. The chief industrial users of gases were iron and steel producers. They made few complaints to the Commission, but the government was interested in increasing the use of oxygen in iron and steel production and the Ministry of Fuel and Power supported such a reference, and their combined weights were sufficient eventually to overcome the objection of the lack of sanctions against large firms.[82] In the case of tyres, dominated by Dunlop, it was a combination of the motor vehicle industry and local authorities and many other public bodies.[83] In rainwater goods the building contractors were anxious for action to reduce the web of restrictions, and were prime movers of the earlier report on building materials.[84] In the case of lino the countervailing power was a few important firms outside the trade association,[85] and in dental supplies it was an alternative trade association.[86]

The reference of semi-manufactures of copper and other non-ferrous metals was the direct and immediate response to a claim from the European Co-operation Administration in Britain that the Lausanne agreement of the BNFMA contravened the terms of the European Recovery Programme.[87] Complaints from Commonwealth governments against heavy electrical equipment and electric wires and cables were an important element also in the referral of two major industries, although in the case of cables the controversial nature of the reference was reduced by specifically omitting the international arrangements from the remit of the Monopolies Commission enquiry.[88]

The calico-printing referral sprang from a long-standing desire to begin to tackle restrictive practices in textiles, especially in the finishing trades, while avoiding taking on the yarn spinners.[89] Some customers of the Federation of Calico Printers were critical of the prices charged and the quality of production.[90] The failed proposal to investigate the yarn spinners had been requested by 'other sections of the industry' who had 'become strongly critical of the spinners' margins, which are important in determining the price of the finished cloth and the markets which can be found for it'.[91]

Obviously, where Ministers and departments raised no objection to an item less discussion occurred, and it is therefore less apparent where the support for the remaining referrals originated. It may be noted, however, that items referred were generally cases of restrictive practices. Indeed, only the case of gases was a straight one-firm monopoly reference, all the rest involved investigation of the large firm as the background to the working of the restrictive practices in the trade concerned.

There was a heavy emphasis on goods used in the construction industry: structural steel, new buildings in London, sand and gravel in central Scotland, rainwater goods, electric street-lighting equipment and even standard metal windows and doors, and imported timber. This may be said to reflect a certain 'public interest' concept operating, for these were cases where the most vocal complaints came from local authorities and other public bodies, and which were important in the current house-building programme. Even here building materials were an area where, in the 1940s at least, there was a strong lobby from building contractors against restrictive practices in distribution.

This assessment of the major features of the choice of referrals indicates that the Board of Trade was hamstrung, if not captured, by fears of antagonising industrial interests. Many of the large items referred were ones whose prices affected the costs of intermediate goods, purchased by other industrialists. Here the outcome could be expected to contribute to competitiveness. However, others were marginal, aimed at gathering information about restrictive practices in general and avoiding acting on controversial industries and sensitive monopolies.

The origins of the two Section 15 references into practices commonly encountered support this picture of governmental timidity. The only Section 15 reference made and completed was that on collective discrimination. Ostensibly it was a sign of Conservative resolve on restrictive practices. Actually it was conceived as a way to avoid taking action on an earlier report, the Lloyd Jacob enquiry into resale price maintenance. This had recommended making certain forms of the practices illegal, and had prompted a tough White Paper from the Labour government making all forms of resale price maintenance illegal. Large-scale lobbying by business interests delayed any legislation by Labour. An early meeting with the new President prompted the adoption of 'no action' and 'a further process of enquiry'. The aim was to avoid a piece of legislation which would 'meet with strong opposition from the trade interests concerned.'[92]

The second Section 15 reference, into common prices and uniform tendering, arose in 1955 and met with strong opposition, especially from the oil companies and the newly-denationalised iron and steel companies. The Board thankfully dropped the idea once it was clear that the 1956 Act was to be introduced.[93]

4 Conclusions

This study indicates that the place to look for signs of regulatory 'capture' in Britain are in the government departments controlling the regulatory body. The power of British governments from the end of the war to act against monopolies and restrictive practices was constrained by business pressure on government departments — Ministers and civil servants — directly. This pressure affected both the nature of the legislation in 1948 and its interpretation during the lifetime of the MRPC.

The ability of sectional interests to immobilise the legislation stemmed from the strategic and structural position which some firms, and certain well-organised trade associations occupied in the economy. The fact that the

Monopolies Commission did not accommodate itself to accusations from industry, but proceeded to produce some highly critical and often very damaging reports, each of which caused a minor public outcry, was to prove its downfall. In the 1956 legislation the Commission was downgraded and there was effectively no regulation of large firms from then until the Monopolies and Mergers legislation.

The question arises whether this failure of governments to resist sectional pressure vitally reduced British competitiveness in the period. In spite of the critical nature of the Commission's reports, both in monopoly and restrictive practices enquiries, and assertions by economists, there is no proof that tougher competition legislation since 1956 has improved British competitiveness and industrial efficiency. Indeed the 1956 Act may have simply encouraged the reassertion of the trend towards concentration and large-firm dominance of markets.[94] One can only speculate that trust- and cartel-busting immediately after the war, instead of ten years later might have given British industry a dynamism to put it in a stronger position to compete with the reconstructed economies in the 1950s. Certainly the persistence of British participation in international cartels into the post-war period seems hardly in line with the emerging expansion of world markets. However, this was an area rarely given to the Commission to investigate, precisely because of the delicate nature of that involvement.

Where government boldness may have had a positive impact would have been in undermining the political strength of business interests through an attack on large firms and trade associations. Forms of business representation to government and hence the structure of government-industry relations changes with developments in the economic organisation of business, especially the growth of large firms and trade associations. A sustained attack on these economic institutions, therefore, especially if well publicised as was the intention, could have made more room for government to manoeuvre in other efficiency-directed policies, such as Development Councils and productivity.[95] That no such attack was mounted is evidence of the rigidities within the British political and economic system.[96]

Notes

1 PRO, BT64/499, H. Wilson, President of the Board of Trade, to Herbert Morrison, Lord President and Stafford Cripps, Lord Chancellor, 3 March 1949, explaining the high salary of the Commission chairman.

2 *Employment Policy*, British Parliamentary Papers (BPP) 1943–44, Cmd. 6527, viii, para. 54.

3 Department of Trade and Industry, *DTI — the Department for Enterprise*, London, 1988, Cm. 278, pp. i–iii.

4 D. Swann, *The Retreat of the State*, 1988; J. Vickers and G. Yarrow, *Privatisation and the Natural Monopolies*, Public Policy Centre, London, 1985; J. Kay and D. Thompson, 'Policy for industry', in R. Dornbusch and R. Layard (eds),

The Performance of the British Economy, Oxford, 1987; T. Sharpe, 'Privatisation, regulation and competition', *Fiscal Studies*, V, 1984.

5 *Meet the Challenge, Make the Change*, Final Report of Labour's Policy Review for the 1990s, London, 1989, pp. 15–16.

6 L. Hannah, *The Rise of the Corporate Economy*, London, 1983, pp. 180–1; J. S. Prais, *The Evolution of Giant Firms in Britain*, Cambridge, 1980, p. 4; P. E. Hart and R. Clark, *Concentration in British Industry, 1935–1975*, London, 1980, pp. 13 and 27.

7 Hannah, *Corporate Economy*, p. 135.

8 For more detail on this whole question see H. Mercer, *The Evolution of British Government Policy towards Competition in Private Industry, 1940–1956*, unpublished Ph.D thesis, University of London, 1989.

9 Mercer, '*Competition*', pp. 51–2; J. D. Gribbin, 'Pre-war cartels in the United Kingdom', in Board of Trade, *Survey*, p. 14; D. C. Elliott and J. D. Gribbin, 'The abolition of cartels and structural changes in the United Kingdom', in A. P. Jacquemin and H. W. de Jong (eds), *Welfare Aspects of Industrial Markets*, Leiden, 1977, pp. 358–60

10 Gribbin, 'Pre-war cartels', p. 14; G. C. Allen, *Monopoly and Restrictive Practices*, London, 1968, p. 92; L. Hannah, *Corporate Economy*, p. 136.

11 *The Economist*, 14 May 1955 and 14 January 1956; *The Times*, 15 June 1955, 18 February 1956, 21 February 1956; *News Chronicle*, 13 May 1955, 8 June 1955, 28 May 1955; MRPC *Report on the Supply of Buildings in the Greater London Area*, BPP 1954, xvi, p. 67 and passim.

12 See especially a key paper written by G. C. Allen and H. Gaitskell in 1943 for the Board of Trade, 'The control of monopoly', PRO, BT64/318.

13 PRO, CAB/87 3, RP (43) 25, 'Report on the recovery of the export markets and promotion of the export trade', 15 June 1943.

14 'Let us face the future', in F. W. S. Craig (ed.), *British General Election Manifestoes 1900–1974*, Chichester, 1982.

15 Hansard, *House of Commons Debates*, vol. 448, col. 2019, 22 April 1948.

16 E. G. Eric Fletcher (Islington East), col. 2059.

17 E. G. Sir A. Plant, 'Monopolies and restrictive practices', *Lloyds Bank Review*, X, 1948.

18 Hansard, *House of Commons Debates*, vol. 549, col. 1927, 6 March 1956.

19 Dalton Diaries, I, Box 31, 4 February 1944.

20 Mercer, *Competition*, pp. 123–8 and 177–80.

21 Hansard, *House of Commons Debates*, vol. 449, col. 2027, 22 April 1948.

22 Mercer, *Competition*, pp. 177–81.

23 W. T. Morgan, 'Britain's election: a debate on nationalisation and cartels', *Political Science Quarterly*, LXI, 1946, pp. 227–33.

24 PRO, BT13/220A, MM (46) 27, 'Restrictive practices', note by the department for the President of the Board of Trade's morning meeting, undated, discused March 1946.

25 A. Rogow and P. Shore, *The Labour Government and British Industry, 1945–1951*, Oxford, 1955, p. 68.

26 Hansard, *House of Commons Debates*, vol. 449, col. 2027, 22 April 1948; PRO, BT64/467, note by R. C. Bryant, Asst. Sec., Board of Trade, 10 March 1948.

27 PRO, BT64/467. Note by Alix Kilroy (Dame Alix Meynell), Asst. Sec., Board

of Trade, and the first Secretary of the MRPC. The note reports on a meeting with other departments about the Bill, 6 March 1948. Although there is evidence of some shift in attitudes by 1948, Mercer, *Competition*, p. 181.

28 Paul H. Guenault and J. M. Jackson, *The Control of Monopoly in the United Kingdom*, 2nd. ed., London, 1974, pp. 22–95; C. K. Rowley, *The British Monopolies Commission*, London, 1966; Allen, *Monopoly*, pp. 70–88; Alex Hunter, 'The Monopolies Commission and economic welfare', *Manchester School of Economic and Social Studies*, XXII, 1955, pp. 22–39; Alix Kilroy, 'The task and methods of the Monopolies Commission', *Transactions of the Manchester Statistical Society*, 1952–53. The exception is an analysis of the Prices Commission which is firmly based in the American regulatory tradition, J. Tomlinson, 'Regulating the capitalist enterprise: the impossible dream?', *Scottish Journal of Political Economy*, XXX, 1983. In addition a study of the Import Duties Advisory Committee (IDAC) engages with the problem, F. Capie, *Depression and Protection: Britain Between the Wars*, London, 1983.

29 Allen, *Monopoly*, p. 88. See also Hunter, 'Monopolies Commission', p. 82; J. Jewkes, 'British monopoly policy 1944–56', *Journal of Law and Economics*, I, 1958, p. 8; J. D. Gribbin, 'Recent anti-trust developments in the UK', *Anti-Trust Bulletin*, XX, 1975, p. 22.

30 See for example, Harold U. Faulkner, *The Decline of Laissez-Faire*, New York, 1951. For a more detailed account of the 'public interest' literature see T. McCraw, 'Regulation in America: a review article', *Business History Review*, IL, 1975, pp. 159–83 and R. Posner, 'Theories of economic regulation', *Bell Journal of Economics and Management Science*, V, 1974, pp. 335–58.

31 Gabriel Kolko, *Railroads and Regulation 1877–1916*, New York, 1965, p. 3 and passim.; George Stigler, *The Citizen and the State*, Chicago, 1975, p. 114; Samuel P. Huntington, 'The Marasmus of the I.C.C.: the commission, the railroads and the public interest', *The Yale Law Journal*, LXI, 1952, p. 470. Gabriel Kolko, *The Triumph of Conservatism*, New York, 1963.

32 M. H. Bernstein, *Regulating Business by Independent Commission*, Princeton, 1955, p. 100.

33 *Sixth Report from the Select Committee on Estimates, session 1952–53: Monopolies and Restrictive Practices Commission*, BPP, 1952–53, IV, pp. 46–7.

34 This was in electric lamps, annual report by the Board of Trade on the work of the Commission: for 1953, BPP, 1953–54, XVI, p. 6 and BPP, 1955–56, XXVI, p. 5; dental goods — report for 1952, BPP, 1952–53, XV, p. 5 and for 1953, p. 6; and timber — report for 1955–56, p. 5 and Monopolies Commission, *Imported Timber: Report on Whether and to What Extent the Recommendations of the Commission Have Been Complied With*, BPP, 1958, XVI.

35 Report for 1955–56, p. 5.

36 Rowley, *Monopolies Commission*, pp. 335–9; PRO, CAB134/846, EA(53) 18th meeting, 17 June 1953.

37 Rowley, *Monopolies Commission*, pp. 353–8; J. B. Heath, 'Restrictive practices and after', *Manchester School of Economic and Social Studies*, XXIX, 1961, p. 199.

38 PRO, CAB134/855, 'Report on the process of calico printing', Memorandum by the President, 2 February 1955; CAB134/846, EA(53) 32nd meeting, 16 December 1953.

39 PRO, BT64/4837, MC/RPM(52)11, 15 July 1952.

40 Allen, *Monopoly*, p. 88; Jewkes, 'Monopoly policy', p. 8.

41 Pilkington Archives, Executive Committee minutes, PB260, 12 January 1949, 7 June 1949, 10 May 1950; General Board minutes, PB350, 26 January 1950, 30 March 1950, 25 May 1950; Working papers for the General Board, PB257, 31 March 1949, 30 November 1950.

42 Pilkington Archives, PB257, 30 November 1950.

43 E.g. Allen, *Monopoly*, p. 88; Jewkes, 'Monopoly policy', p. 8; Hunter, *Competition and the Law*, pp. 82–3.

44 Mercer, *Competition*, pp. 272–320. See also R. B. Stevens and B. S. Yamey, *The Restrictive Practices Court*, London, 1965, pp. 15–16.

45 D. Maxwell-Fyfe, *Political Adventure*, London, 1964, p. 261

46 PRO, BT64/4996, Letter from D. Walker, Chairman of BEAMA, to the President, 11 May 1953, and note of meeting between the Board of Trade and BEAMA, 1 February 1954.

47 Hansard, *House of Commons Debates*, vol. 537, col. 1578, 24 February 1955.

48 Jewkes, 'Monopoly policy', p. 13.

49 Hunter, *Competition and the Law*, p. 25; Allen, *Monopoly*, p. 80; Rowley, *Monopolies Commission*, p. 193.

50 Allen, *Monopoly*, p. 80; Hunter, *Competition and the Law*, p. 82; Jewkes, 'Monopoly policy', p. 4.

51 Huntington, 'The I.C.C.', p. 473.

52 PRO, BT64/4824, file on excisions from the tyres report; BT64/4854, file on excisions from the report on lino; BT64/4879, files on excisions from the report on gases; BT64/4882, file on excisions from the report on electrical machinery; BT64/4889, file on excisions from the report on tea; BT64/4890, file on excisions from the report on metal windows; BT64/5003, file on excisions from the report on London buildings; BT64/4852, file on excisions from the report on copper semi-manufactures. BT64/4890 contains a summary to excisions in reports up to 1956.

53 PRO, BT258/71, meetings between Board of Trade and representatives of Dunlop, 15 August 1955 and 6 September 1955; BT258/72, meeting at the Board of Trade with the President 19 January 1956; BT64/4852, paper by the British Non-Ferrous Metals Federation; letter to A. C. Hill, 19 January 1956.

54 PRO, BT64/4837, MC/RPM (52)11 and 12, 15 July 1952 and 17 July 1952, papers of the Lee policy group on monopolies.

55 PRO, BT64/4812, 'Monopolies and Restrictive Practices Commission, selection of references 5th group', letter from Monopolies branch of the Board of Trade to Board of Trade production departments, 26 September 1953.

56 PRO, BT64/4813, letters A. C. Hill to Sir Maurice Dean (2nd Sec., Board of Trade), 24 July 1954 and 5 August 1954. BT64/4887, 'The work of the Monopolies Commission,' note by A. C. Hill, September 1954.

57 *Ibid.*; Hill was supported by the Assistant Sec., BT64/4754, Note by T. K. Rees on Ministry of Supply cases, 22 March 1955.

58 PRO, CAB134/854, EA(55)28, 'Policy on restrictive practices' Memorandum by the President of the Board of Trade, 7 February 1955, and EA(55), 6th meeting, 17 February 1955; CAB128/29, CM(55), 12th meeting, 7 June 1955 and CM (55), 15th meeting, 16 June 1955.

59 Jewkes, 'Monopoly policy', p. 7; G. Polyani, *Which Way Monopoly Policy?*, IEA Research Monograph 30, 1973, p. 63 speculates on this question for a later period.

60 Evidence which follows on the Board's screening procedure is all drawn from long and tedious discussions in a few files: PRO, BT64/2757, possible cases for enquiry, discussions on the 1949 referrals; BT64/4833, 4834, 4835, 4836, 4837, papers of Sir Frank Lee's policy group on the Monopolies Commission, 1951–52 BT64/4812 and 4813 — selection of references for the 5th group; BT64/4754 — consideration of items for referral 1954–55; BT64/4838 on the future of the Commission, 1955–56.

61 PRO, BT64/2757, unsigned circular letter to Board of Trade production departments, March 1948; BT64/4837, MC/RPM(52)12, 'Further references to the Monopolies Commission', paper to the policy group on the MRPC, 17 July 1952.

62 BPP, *Report of the Select Committee*, p. 18.

63 PRO, BT64/4812, letter from A. C. Hill to Ministries of Works, Materials and Supply, 27 October 1953.

64 This was an argument used against the referral of razor blades, industrial gases, metal cans and containers, gas mantles, potash and footwear repairs, all under the control of one or a few large firms. Potash was an international cartel run jointly for the British side by ICI and Fisons.

65 PRO, BT64/4812, 'MRPC selection of references: fifth group', note from Monopolies branch of the Board of Trade to production departments, 16 September 1953.

66 PRO, BT64/4837, papers of the Policy group on new legislation concerning the Monopolies Commission and resale price maintenance, MC/RPM(52)12, 17 July 1952.

67 PRO, BT64/4812, minutes 21 September 1953 and 13 October 1953, and undated note by R. Colegate, Asst. Sec.

68 PRO, BT64/2757, 'MRPC possible cases for enquiry', minutes of meeting at the Board of Trade, 12 December 1948. Other cases are heavy electrical machinery, diamonds, metal cans and containers.

69 PRO, BT64/4754, letter Geoffrey Lloyd, Minister of Fuel and Power, to Thorneycroft 15 March 1955, letter Gerald Reading, Minister of State at the Foreign Office, to Thorneycroft, 17 March 1955; minutes of meeting at the Board of Trade, 15 March 1955.

70 PRO, BT64/4813, 'Suggestions made in Parliamentary questions for reference of subjects to the Monopolies Commission'. This showed that between 1952 and 1954 there had been twelve Parliamentary Questions on prices and other matters in the oil and petroleum industries.

71 *Royal Commission on the Press 1947–1949. Report*, BPP, 1948–49, XX, Cmd. 7700, Appendix IV.

72 PRO, BT64/4837, MC/RPM (52)12, 'Further references to the Monopolies Commission', 17 July 1952, and BT64/2757, minutes of meeting 12 December 1948.

73 PRO, BT64/4813, 'Possible subjects for reference to the MRPC' fifth group.

74 W. J. Reader, *Imperial Chemical Industries — a History: Volume II*, Oxford, 1975, pp. 475–7.

75 PRO, BT64/2757, official in Board of Trade Raw Materials Dept. to R. C. Bryant (Asst. Sec., Board of Trade), 14 April 1948.

76 PRO, BT64/2757, goods considered for shortlist, 1948.

77 PRO, BT64/4812, letter to D. R. McGregor (Under Sec., Board of Trade) to A. C. Hill, 30 October 1953.

78 PRO, BT64/4754, E. Harwood to A. C. Hill, 12 March 1955.

79 PRO, BT64/4754, minutes of meeting at the Board of Trade, 3 March 1955.

80 PRO, BT64/4821, letter from A. C. Hill to the Board of Trade Solicitor, 25 June 1955, on the views of the iron and steel division of the Ministry of Supply. BT64/4826, file generally on the Iron and Steel Board's comments on the Section 15 reference.

81 PRO, CAB134/844, EA(52)108, 'Monopolies and restrictive practices: selection of a new reference to the Monopolies Commission', note by President, 25 July 1952.

82 PRO, BT64/4837, MC/RPM(52)12, 'Further references to the Monopolies Commission', 17 July 1952; BT64/2790, report of a meeting by A. D. Neale, 25 November 1949; MRPC, *Industrial and Medical Gases*, BPP, 1956–57, XVII, p. 40.

83 PRO, BT258/70, R. R. D. McIntosh to A. E. Lee (Asst. Sec., Board of Trade), 24 September 1948 and Alix Kilroy to McIntosh, 22 October 1948, expressing the desire to refer the industry in spite of industrial opposition.

84 Ministry of Works, *The Distribution of Building Materials and Components*, HMSO, 1948, p. 25; Hansard, House of Commons Debates, 419, 11 February 1946, written answer, 28; J. Jewkes, *Ordeal by Planning*, London, 1948, p. 57.

85 MRPC, *Report on the Supply of Linoleum*, BPP, 1955–56, XXIV, p. 17. Their position and the opposition of some wholesalers to the Linoleum Manufacturers' Association schemes may have been less important as a cause of referral than complaints by local authorities, p. 28.

86 MRPC, *Report on the Supply of Dental Goods*, BPP, 1950–51, XVIII.

87 PRO, BT64/4852, note on the background to the copper semis reference, 24 October 1955; BT64/502, exchange between the Ministry of Supply and the Board of Trade, Dec. 1948–Feb. 1949 on a complaint also from the New Zealand High Commission.

88 PRO, BT64/2757, 'Programme for the first year of the MRPC — shortlist', 21 October 1948.

89 PRO, BT64/2757, 'Programme for the first year of the Monopolies Commission', 21 October 1948.

90 MRPC, *Report on the Process of Calico-Printing*, BPP, 1953–54, XVI, pp. 57–8.

91 PRO, CAB134/844, EA(52)108, 'Monopolies and restrictive practices: selection of a new reference to the Monopolies Commission', note by President, 25 July 1952.

92 PRO, BT64/4887, paper by Monopolies branch, 7 December 1951 and MC/RPM(51)22, 8 December 1951.

93 PRO, BT64/4821 and 4822, files generally on oil lobbying and the reference.

94 Elliott and Gribbin, 'Abolition of cartels', pp. 357–8.

95 In the former, the political opposition which trade associations and the FBI mounted killed the Labour government's initiative. Rogow and Shore, *Labour Government*, p. 82.

96 On this see P. Hall, 'The state and economic decline', in B. Elbaum and W. Lazonick (eds), *The Decline of the British Economy*, Oxford, 1986.

I would like to thank Mike Collins and Jim Tomlinson for comments on an earlier draft of this chapter, and also Messrs Pilkington for permission to cite their archives.

6 *Wyn Grant*

Government and manufacturing industry since 1900

The apparent strength of British manufacturing industry in the nineteenth century was based largely on a narrow range of staple industries such as coal and textiles, and the absence of effective foreign competition. As Kirby and Rose point out in their chapter, political awareness of Britain's industrial vulnerability developed in the late 1880s and 1890s, but prevailing economic and political orthodoxies prevented the development of an effective response by government. This continued to be the case for much of the twentieth century, so that a review of British industrial performance published by the National Economic Development Office in 1980 concluded, 'In every respect the relative achievements of the UK have been, and remain, poor.'[1] Improvements on some economic indicators in the 1980s may conceal persistent underlying problems. At the end of the decade there was a severe and persistent deficit on manufacturing trade, while even the surplus on 'invisibles' went into deficit for the first time. In 1989, productivity began to fall, earnings continued to run ahead of inflation, and manufacturing output went into decline.

Throughout the twentieth century, British manufacturing industry has been characterised by the slow and painful decline of the old staple industries, and their slow and hesitant replacement by new industries, whether the electrical industry before and after the First World War, or information technology in the 1970s and 1980s. The staple industries continued to enjoy a political displacement out of proportion to their long-term economic significance, leading government to nurture them with protection and cartels in the inter-war period, and subsidised public ownership in the post-war period. In its dealings with industries such as coal, government showed a persistent reluctance throughout the inter-war period to intervene effectively to deal with what were generally recognised as problems of inefficient structure and poor management. As Chick notes in his chapter, even when government decided to nationalise basic industries, few of the supposed benefits of nationalisation such as improved investment decision-making and sectoral co-ordination were secured in the early post-war period.

Although 'Britain's economic record since 1945 represents a notable improvement on the interwar period ... increasingly it fell well short of current expectations and the achievements of our major competitors.'[2] The central problem was 'the failure of British industry to maintain its share of world trade in manufactures, on the one hand, and on increasing import penetration of the domestic British market, on the other.'[3] Since 1963 imports of manufactures as a share of gross domestic product have risen at a trend rate of 6 per cent a year, but exports of manufactures as a share of GDP have grown at only 2 per cent.[4] The share of manufacturing in GDP in the UK is lower than in any other major industrial country: 22 per cent of GDP in Britain compared with 29 per cent and 33 per cent respectively in Japan and West Germany.[5]

During the post-war period, government attempted to take a more systematic interest in the problems of manufacturing industry, but much analysis and diagnosis rarely led to well thought out or systematically implemented policy measures. Poor policy performance was not accidental, or the result of incompetence on the part of individual politicians or civil servants. Unlike countries which have followed up a 'catching up' strategy, such as Germany, the fact that industrialisation occurred in Britain without, and before the existence of, a modern state would appear to have created a deep ambivalence about the role of government in relation to industry. There has been a continual tension between those who advocate a *dirigiste* approach to Britain's industrial problems, and those who favour liberal solutions. This tension has been particularly apparent in the country's principal governing party, the Conservative Party, with the tide of opinion swinging backwards and forwards between the etatiste and liberal tendencies.[6]

In the 1980s, it was, of course, the liberal tendency under Mrs Thatcher which set the tone of policy. Just how recent a development this is is apparent when one examines a now forgotten, but at the time significant, document published in 1976, *The Right Approach: a Statement of Conservative Aims*. After referring to the desirability of a mixed economy, the document goes on to comment, 'The precise limits that should be placed on intervention by the state are reasonably the subject of debate within the Conservative Party, as are the proper boundaries between State and private provision.'[7]

No such uncertainties have troubled the policy path of the Thatcher government. It could be argued that the government has implicitly favoured service industries, particularly financial services; manufacturing industries have been left to sink or swim. The government would, of course, point to a number of improvements in the performance of manufacturing industry in the 1980s. Profitability has recovered from disastrously low levels at the beginning of the 1980s, reaching the highest level for twenty years in 1988. Labour productivity, for a long time one of the areas of greatest weakness of British industry, for a time increased at a rapid rate by British and international standards. Manufacturing output per head rose at an annual average percentage rate of 5.25 per cent between 1980 and 1988 compared with 1.5 per cent between 1970 and 1980. Performance towards the end of the 1980s was less impressive, with output per head across the economy as a whole starting to fall in 1989.

Moreover, there are many reasons for concern about industrial performance. The output of manufacturing rose by only 12 per cent between 1979 and the first quarter of 1989. Unit labour costs in manufacturing rose more rapidly in the UK than in other major industrial countries in 1988 and 1989. Manufacturers continue to refer to skilled labour shortages as a significant constraint in CBI surveys. The results of the most extensive survey into the state of vocational training in Britain were described as 'mind boggling' by the Secretary of State for Employment after it was found that about two-thirds of employees questioned said that they had not undergone any training in the past three years, and more than 40 per cent could not imagine themselves undertaking any training.[8] According to OECD data, Britain was the only western industrial country in which the percentage of national income devoted to research and development declined during the first half of the 1980s.[9]

Analysts who consider that the economy has made significant advances in the 1980s nevertheless identify inadequate training and insufficient investment in research and development as two key weaknesses of the British industrial economy.[10] These weaknesses are of particular significance given the transformations being experienced by economies throughout the world. The nature and extent of these changes has been the focus of the debate about a shift from a Fordist to a post-Fordist mode of production and consumption. It is not possible to discuss this far-reaching debate in any depth here. In summary, what is argued is that Fordist mass production and consumption is being displaced by a new mode of 'flexible specialisation.'[11] Instead of long production runs aimed at serving mass markets, the potentialities of computer controlled production technology will be utilised to produce high quality goods aimed at particular niches in the market.

Standardised, mass production goods will, of course, continue to be required, but these needs will be largely met by newly industrialising countries which are highly competitive on labour costs. Advanced industrial countries will, it is predicted, succeed if they are able to bring together advanced production technology, good design, and an ability to respond quickly to changes in consumer tastes. Another implication of these trends is that industrial success in the twenty-first century will depend to a considerable extent on the ability to fuse knowledge-intensive services with the manufacturing process. In other words, the political debate about the relative importance of manufacturing and services, which arguably has distracted British policy-makers from more fundamental problems, may cease to be relevant.

1 The burden of history

In the last decade of the century, it is tempting to look forward to the future, yet one cannot properly understand the British predicament without analysing the burden of the past. One of the most disheartening aspects of the debate about British industrial competitiveness is the extent to which problems highlighted by analysts at the end of the nineteenth century are still with us at the end of the twentieth century. In his 1896 analysis of the problems of the British industrial economy, E. E. Williams dismissed the arguments that excessive strikes or high wages were causes of Britain's declining industrial supremacy. He argued that the importance of strikes could be overrated, and pointed out that German wages were higher, and hours shorter, than in Britain.

Instead, he focused on such factors as the low productivity of British manufacturing plants, especially compared with those of Germany. This very deficiency has been the focus of a series of recent studies by the National Institute for Economic and Social Research in which they have compared plants producing matched products in Britain and Germany, attributing the marked differences in productivity to such factors as differences in standards of training and poor organisation of the flow of work on the factory floor.[12] Williams also noted the greater attention to scientific and technical education in Germany; the insufficient priority given to commercial relations by the diplomatic service; the poor record of English exporters in learning foreign languages; and the lack of willingness of English firms to introduce variations in form, design and quality to meet customers' needs (an observation highly relevant to the current flexible specialisation debate). His solutions were more spending on education, and particularly technical education; more attention should be paid to the specific needs of customers; language education should be improved; manufacturing plant should be updated; and a greater priority should be given to improved design.[13] This agenda for action remains relevant today.[14]

Although the need for an historical analysis of government and the competitiveness problem in Britain should be evident, the existence of important divergences of interpretation should not be underestimated. For example, the Thatcher government's explanation of poor British economic performance emphasised corporatism and excessive trade union power:

> The ability of the economy to change and adapt was hampered by the combination of corporatism and powerful unions. Corporatism limited competition and the birth

of new firms whilst, at the same time, encouraging protectionism and restrictions designed to help existing firms. Trade unions opposed changes in working practices and fed the inflationary expectations of their members.[15]

An alternative account would be that the real problems were a failure to develop an effective working relationship between government and industry, and the presence at the firm level of implicit collusions between slack managements and conservative trade unions to preserve the *status quo* for their mutual benefit. Certainly, most analysts would argue that Britain has not been marked by the presence of corporatism, but by its absence. Admittedly, Middlemas has argued that the British system 'is not corporatism, but one where *corporate bias* predominates.'[16] His comparison of corporate bias to 'the bias of a wood at bowls'[17] is compelling and relevant, but he overlooks the fact that the corporatist bowls game in Britain has been played on a green built on a sub-structure of individualistic, liberal assumptions. Hence, the wood can veer in entirely unexpected directions.

There have been a number of attempts to develop corporatist relationships in Britain (for example, after the First World War and in the 1960s and the 1970s), but they have ended in failure. One of the reasons for their failure has been that they have cut across the prevalent liberal, voluntaristic paradigm in Britain. Corporatism requires effective business associations and trade unions, and the presence of such bodies is dependent in part on state intervention, which may be acceptable in a country engaged in a 'catching up' strategy such as Germany, but has not been acceptable in Britain even in wartime conditions.[18]

This is not to suggest that Britain should have adopted a corporatist strategy. It may be argued that the problems of government-industry relations are more fundamental than can be resolved by constructing a new organisational relationship between government, employers and unions. In part, the problem is one of values, the persistence of a value system which emphasises the autonomy of the individual entrepreneur rather than co-operation between government and industry in the pursuit of long-run strategic objectives on the Japanese model. Another fundamental problem is the nature of the British civil service which, for all its many strengths, is not suited to forging an effective relationship with industry.

Although there has been some loss of promising young civil servants, particularly to the financial services sector, in the 1980s, and although provision for secondments in both directions has increased, civil servants and industrial managers follow separate, lifetime career paths. There is no equivalent of the Japanese practice of 'descent from heaven' which leads to the large scale transfer of senior civil servants into industry,[19] unless one counts the recruitment of retired senior civil servants from the Ministry of Defence by defence contractors. Britain retains a generalist civil service which is highly skilled at briefing ministers, preparing them for parliamentary questions and appearances before select committees, and at keeping the government machine functioning in periods of crisis. It is, however, a civil service which has only limited knowledge of how industry works, and in which, despite the hopes raised at the time of the Fulton Report, relevant training is on a very limited scale.[20]

As far as the question of trade union power is concerned, there is evidence from a number of industries that the real problem is collusion between management and unions to perpetuate inefficient practices in the pursuit of a quiet life. Admittedly, the pressures of the 1980s have brought some changes in this area, although how enduring they are remains to be seen. Certainly, some of the structural changes which have taken place, such as union mergers, may enhance the potential of inter-union conflict, given that they have not generally produced one plant unions, but have often altered the balance of power between different groups of workers in the workplace (e.g. non-manual workers in MSF may outnumber any group of organised manual workers).

Lazonick's study of the cotton industry shows how individualistic managers in family-based firms adopted an industrial relations strategy, which although it avoided competition on labour costs with other firms and continual conflict with the workforce, reduced the industry's ability to adapt to changing conditions. Far-reaching rights were conceded by the management to the workers:

> Well into the twentieth century, British cotton firms were constrained to abide by well worked-out and deeply entrenched wage lists that had their origins in the last half of the nineteenth century. In the 1920s and beyond, the British cotton industry was afflicted by a severe case of institutional rigidity: it could not adapt organizational structures that had developed during its period of international dominance to the new requirements of international competition.[21]

Similar problems were apparent in the engineering industry in Coventry in the post-war period. At the Standard plant, the 'big gang' system represented a highly developed form of shop-floor control, with union control of recruitment. Considerable internal friction between the gangs was one consequence, focused particularly on competition for lucrative pieces of work.[22] At the Rootes plant, there was a style of 'management by abdication'. Management regularly gave in on small issues to keep production going ... At shop floor level, little attention was given to supervision which was poorly paid and had little or erratic backing from senior management.'[23]

In all these cases, the style of labour-management relations was the result of management decisions, plus the willingness and ability of the unions to seize the opportunities for shop-floor control which were presented to them. Government did not influence the bargains struck. As industrial relations increasingly became seen as a central problem of the British economy, government started to take a more systematic interest in its conduct, over and beyond the traditional conciliation functions performed by the Ministry of Labour. However, this effort created very serious political problems for first the Wilson government and then for the Heath government. Only against a background of high unemployment, and large-scale structural change in the economy, was it possible to limit the influence of the unions in the 1980s. Even then, the British problem of earnings running ahead of inflation and productivity gains seemed as intractable as ever.

Peter Hall has argued that British economic decline can be explained in terms of a particular institutional configuration, one feature of which was political inertia. He states that 'The inertia of the political system became most important during the critical conjunctures that followed the two world wars.'[24] In other European countries, the experience of intervention during the wartime period, and the dislocation at the end of it, encouraged a change of direction in government–industry relations. Why were fundamental changes not experienced in Britain?

Hall himself places considerable emphasis on the personality and political reputation of Lloyd George. Whilst this was undoubtedly a factor, other forces were at work. A group of 'productioneers' centred around the founder of the Federation of British Industries, Dudley Docker, were demanding major changes in government–industry relations. They wanted merger and rationalisation, a share for labour in the benefits of success and, most significantly from the perspective of this chapter, 'they wanted government, industry and finance to co-operate in rationalisation and export policy'.[25] They certainly had no time for Lloyd George's political opportunism. However, as Kirby and Rose show in their chapter, the military collapse of Germany undermined the arguments of the advocates of industrial reconstruction.

That is not to say that the experience of government–industry co-operation during the First World War had no lasting beneficial consequences. There were a number of useful initiatives taken during the inter-war period which were facilitated by the existence of a network of personal contacts, largely arising from the wartime period, between industrialists, financiers, politicians and civil servants. This informal network 'co-operated to promote, but also control, state intervention in the economy.'[26]

> Through this group British business between the wars was partly reconciled to state intervention in industry. In that way, much more than in its direct effect on the proportion of GNP accounted for by government spending, or in the models it threw up of interventionist institutions, the First World War helped to change the relationship between British industry and the state.[27]

One example of the efficacy of these informal contacts in bringing about significant changes was the formation of Imperial Chemical Industries.[28] This was seen by government as a 'chosen instrument' to transform the

British chemical industry which, despite a systematic effort at industrial espionage against the defeated Germans, was in a dire state in the 1920s. ICI's formation can be traced back to a lunch in January 1926 between Sir Harry McGowan and the chairman of Midland Bank, Richard McKenna, who had been Chancellor of the Exchequer in Asquith's War Cabinet 'and still maintained close links with government.'[29] McKenna suggested a British equivalent of the German chemical combine, IG Farben. ICI subsequently became something of a flagship of the British industrial economy, ranking as one of the five leading world chemical companies in terms of output in the post-war period.[30]

Of all the government interventions in the post-war period which assisted competitiveness, perhaps the most important was the formation of the Central Electricity Board which constructed the national grid. By accepting power from generators on a 'merit order' basis, the CEB forced the least efficient generators out of business, and ensured that new stations (such as that at Battersea in London) were constructed in such a way as to maximise efficiency of output, given the technological knowledge and fuel costs of the period. Not only did the CEB ensure that energy-intensive industries had an efficient power supply, but the falling real cost of electricity stimulated a new industry supplying domestic electrical appliances.[31]

Progress was also made on the basis of government interventions in other industries such as food processing (milk marketing, sugar beet processing).[32] The Bank of England set up the Securities Management Trust in 1928 to deal with its industrial holdings, followed by the Bankers' Industrial Trust with a wider remit of industrial rationalisation. Bank intervention was considerably influenced by the personal whims of the Governor, Montagu Norman, and hopes of a new era in the financing of British industry were quickly dashed:

> In the event these agencies proved not to be the engines of rationalisation through a great array of industries; their principal task was rather to oversee the unwinding of commitments into which the Bank had already ventured, or into which it was about to be driven by political pressures ... little emerged in the way of firm schemes of any substance.[33]

There was more government activity with the implicit or explicit objective of improving competitiveness in this period than was often recognised. Although its scope was limited, it was able to ensure more enduring successes (ICI, the London Passenger Transport Board, the promotion of civil aviation, the milk marketing boards, the National Grid) than the more extensive interventions of the post-war period. Even so, the limits of the *ad hoc*, reactive interventions of the period must be recognised. Allen argues that the predominance of the old staple industries in Britain, and the weakness of newer industries, was reinforced by government intervention to suppress competition through cartelisation, and the supplementation of a general protective tariff by quantitative restrictions on imports.[34]

In the case of the iron and steel industry, 'the National Government was committed to fundamental reorganization but found that the tariff, once enacted, enabled the industry to dictate the course of events'.[35] The outcome was that 'the state sponsored a cartel over which it had little control.'[36] In practice the government 'helped to reinforce old structures with a new layer of powerful institutions that embodied existing structural problems rather than counteracting them.'[37]

Effective intervention to improve competitiveness depends on the organisational effectiveness of both government and industry, facilitating a mature and sophisticated dialogue between them. On the side of government, it was admitted that it 'was not equipped with the administrative and managerial skills to take an active part in the management and reorganisation of industry.'[38] Apart from the CEB and the Board of Trade's Mines Department, 'there was little development of traditions of informed state supervision or direction, nor was there much progress in developing institutions within which government and business could interact to formulate mutually advantageous compromises.'[39] Matters were no better on the part of industry, despite the formation of the Federation of British Industries during the First World War. Turner summarises the relationship between business and the state as 'not one of domination by either side, but of bargaining between two weak entities which often did not know their own minds.'[40] Against this background, it is not surprising that a survey of industrial policy in the inter-war period finds it 'impossible ... to offer anything but an agnostic conclusion on the "efficiency" effects of state intervention in industry.'[41]

2 The Second World War and its aftermath

The Second World War saw a massive expansion of government's capability to intervene in industry, with new and expanded departments staffed to a large extent by recruits from business. On the side of industry, a considerable number of new trade associations were formed, although their survival after the war often served to create further confusion in a structure already characterised by overlap and duplication. Reading association archives of the period, one is often struck that the question 'don't you know there's a war on?' could well have been asked of many of the trade associations which often sought to protect the privileges of their members, resisting compulsion in favour of voluntary run cartel-like arrangements. Barnett, of course, argues that the performance of British industry was characterised by massive inefficiency compared to that of Germany in such vital areas as aircraft production, masked only by massive imports from the United States. Britain's technological breakthroughs such as radar simply served to reinforce the feeling of complacency about Britain's performance felt at the successful conclusion of the war.[42]

The scope of Britain's economic problem was recognised by the Labour government, and they did make a serious effort to tackle the underlying problems of inefficiency through the work of the Dollar Exports Board and the Anglo-American Council on Productivity (AACP). The AACP's work, which emphasised managerial responsibility for poor performance by British industry, is fully discussed in Tomlinson's chapter. The reports from the Anglo-American Productivity Council on particular industries often succeeded in identifying the problems that beset them throughout the post-war period, but as was so often the case in Britain, useful analysis was not accompanied by appropriate action.[43]

The major innovation of the post-war period was, of course, the nationalisation of public utilities and other basic industries such as steel, a topic addressed more fully in Chick's chapter in which he notes that government's role in relation to the industries was afflicted by asymmetries of information and expertise. Even a Conservative Government returning to office would have had to take some action to deal with the problems of run-down industries of vital importance to the national economy, but unable to offer an adequate return to private investors. The way in which nationalisation was carried out is now, however, generally recognised as a failure. The issue of inadequate pricing rules which led to a misallocation of resources in the economy as a whole eventually started to be tackled in the 1960s, but the problem of government's use of the industries as a means of fulfilling its macro-economic objectives was never solved. It damaged the industries by making long-term planning difficult, and it demoralised their managers, particularly after prices and incomes policies undermined industrial relations.

Although the economy was troubled by cyclical problems in the 1950s, and by the ever-present problem of inflationary wage settlements, the extent of Britain's industrial problems was masked by the particular circumstances of the period. Reconstruction, rearmament and population growth sustained demand, whilst the process of recovery on the European continent meant that British exports faced less competition than they would in subsequent decades. Commonwealth markets remained important, accounting for 39.8 per cent of imports and 40.6 per cent of exports in 1955, more if South Africa is included.[44] As the country emerged from the austerity of the immediate post-war years, symbolised by the end of rationing, the emergence of what was hailed as a new 'Elizabethan age', and the wider availability of consumer goods, all helped to engender a mood of some complacency. Nowhere was this more apparent than in prevalent indifferent or negative attitudes towards the developments in continental Europe which were to lead to the formation of the European Community.

Towards the end of the decade it became increasingly apparent that

Britain was lagging behind its major European competitors. Increasing concern in establishment circles led to what has been called the 'Brighton Revolution' of 1960. This referred to a conference organised by the Federation of British Industries at Brighton in 1960 which brought together leading industrialists (including the chairmen of nationalised industries) and senior civil servants to discuss the problems facing the British economy. The significance of this meeting is that it represented a new willingness to consider a more planned approach to the British economy, with the French model attracting some interest, although it is arguable that the extent of intervention in France was not understood.[45] A number of practical consequences followed, including Britain's first attempt to join the European Community, and the formation of the National Economic Development Council as, in effect, an alternative source of economic advice to the Treasury.

With the benefit of hindsight, it is possible to argue that what Brighton ushered in was a period characterised, to use a term coined by Michael Shanks, of 'institutionitis'. The way in which a problem was tackled was to set up a new institution to deal with it. The high point, or low point, of this tendency was the establishment of the Department of Economic Affairs (DEA) by the incoming Labour Government in 1964. Although in some ways its main purpose was to find a safe ministerial home for the mercurial George Brown, it was also seen as providing an institutional counterweight to the economic orthodoxy of the Treasury. It was an institutional battle which the Treasury soon won, leading to the eventual disappearance of the DEA.

Of somewhat greater significance was the formation of the Ministry of Technology, which by the end of the 1964–70 Labour Government gave Britain an industry ministry with wide-ranging responsibilities, and a presumption of the virtues of a technocratic partnership with industry which contrasted with the *laissez-faire* orientation of the Board of Trade. Edward Heath replaced the Ministry of Technology by the Department of Trade and Industry, which was split up by Harold Wilson in 1974, only to be recombined by Margaret Thatcher in 1983. By that time, however, it was a much less significant department, and became even less so as the 1980s progressed, its budget being overtaken by that of the Ministry of Agriculture in 1989. Far from institutional changes leading to more effective policies to promote competitiveness, as was optimistically anticipated in the early 1960s, institutional instability has both reflected and reinforced the incoherence of the policies of successive British governments towards industry.

3 1972 to 1979: the government response to crisis

The period from 1972 to 1979 is of particular importance in the analysis of the relationship between government policy and British industrial competitiveness. It was a period of acute crisis for British industry, symbolised by the collapse of major firms such as Rolls Royce (in 1971) and British Leyland. It was also the period when, starting with the 1972 Industry Act, government made its most sustained attempt to encourage the restructuring of British industry so as to make it more competitive.

The results were not impressive. What emerged was a series of often ill thought out reactive policies which often had the effect of delaying necessary changes. In all fairness to decision-makers, the background against which policy decisions had to be taken was a very unfavourable one. The first oil shock, and the miners' strike of 1973–74, leading to the general election of February 1974, ushered in the period of greatest economic and political crisis Britain has experienced in the post-war period. Against a background of hyper-inflation and industrial unrest, there was talk of the formation of 'private armies' and speculation about the possibility of a military coup. Even so, the best that can be said is that policy-makers managed to muddle through the crisis and avoid total catastrophe.

One of the characteristics of the period was an unwillingness to face up to the seriousness of the problems that the British economy faced, and the drastic nature of the remedies required. This is evident if one examines the cases of the motor industry and the steel industry which absorbed large sums of public money over this period. The 1975 Ryder Report on British Leyland was based on the hopelessly unrealistic assumption that British Leyland would retain a one-third share of the British market and would increase European sales by more than a quarter. These assumptions made it unecessary to propose large-scale redundancies or plant closures. In the case of Chrysler UK, 'A reluctant government was manoeuvred into granting large sums of money to a concern of dubious economic viability for a mixture of economic and party-political motives',[46] although in retrospect the aid given to Chrysler looks somewhat more cost-effective than that given to British Leyland.

Policy towards the steel industry was, if anything, even more of a disaster. The basic problem was that substantial sums of new investment had been pumped into the industry at a time when production was being expanded globally against a background of faltering demand. An implication of the construction of new plants was that some of the older and less efficient plants would have to close, but this was politically difficult to achieve. Lord Beswick's review of proposed steel plant closures, completed in 1975, represented 'an attempt to reconcile the rationalisation and reconstruction of a major industry, with the need to maintain employment in already depressed areas such as Scotland and Wales'.[47] The steel industry's troubles continued, and by 1981–82 taxpayers were paying £50 to £60 a head to support British Steel.[48]

By the time of its privatisation, British Steel's performance had been transformed, albeit at much lower levels of output. By 1988, Britain's pre-tax steel production costs at $415 per tonne were slightly lower than those of newly industrialising countries, and well below other advanced industrial countries such as Japan and West Germany. Between 1980 and 1988, the number of British Steel employees engaged in steelmaking had been cut by 58 per cent while liquid steel output had risen by 24 per cent. Man hours per tonne of steel dropped from over fourteen in 1980–81 to five in 1988. Even so, some persisting competitiveness problems are indicated by the fact that British Steel's productivity was outpaced by Hoeschst of West Germany, while Japanese producers are heading towards three man hours per tonne of steel.[49]

An attempt was made after 1975 to systematise government industrial policy through the Labour government's industrial strategy. This had two major aspects. One aspect, potentially the more important, was to encourage other government departments such as education to give a higher priority to the needs of industry in formulating their policies. The second aspect was the formation of tripartite working parties under the auspices of the National Economic Development Council (NEDC) to discuss and identify the problems of particular sectors of the economy. One difficulty that these bodies faced was that the National Enterprise Board was also pursuing its own industrial strategy in hi-tech sectors of the economy, and was reluctant to give what it regarded as commercially confidential information about its own companies to NEDC committees. Schemes of assistance to encourage investment and rationalisation in particular sectors were set up by the government, sometimes influenced by the discussions of NEDC committees. However, subsequent studies of these schemes showed that they sometimes slowed down rationalisation by providing funds which allowed marginal firms to stay in business.[50]

These difficulties reflected a wider problem with government assistance to industry. Providing financial assistance to firms did not really enable government to intervene effectively in the investment decision at the firm level. Firms were willing to accept government assistance, and indeed some of them became skilled at funding large portions of their investment programmes from government assistance. However, the money was often regarded as a bonus and (with the exception of location decisions) firms were reluctant to allow the availability of government assistance to be the deciding factor in investment decisions.[51] Hence, government efforts to influence competitiveness where having a limited impact where it really mattered, at the level of the firm.

4 Government retreats from the economy

Since 1979 government has retreated from the involvement with the industrial economy which characterised the period from 1972 to 1979, which itself was the culmination of a period of increasing intervention dating back to 1960. The Thatcher government made its position about the role of government in promoting competitiveness clear:

> The central theme for our policies remains the belief that sensible economic decisions are best taken by those competing in the market place. The responsibility of Government is to create the right climate so that markets work better and to encourage enterprise.[52]

It is certainly the case that government assistance to industry has declined significantly under the Thatcher government. A European Community survey shows a 'downward trend with the 1986 aid total about two thirds of the 1981 amount.'[53] This contrasted with an upward trend in Italy and Germany, and an absence of definite trends in other major member states. Since these figures were compiled, there have been further cutbacks in government assistance to industry in Britain. Another indicator of government's declining involvement with industry was the abolition in 1988 of the sponsorship divisions within the Department of Trade and Industry which had been responsible for maintaining relationships with particular industrial sectors.

Although government financial assistance to industry has declined, government has become more involved with industry as a regulator. In part, this regulatory role has developed because nationalised industries have often been privatised in a way which has minimised rather than maximised opportunities for competition, in response to pressure from the managements of the industries themselves, and in the hope of making them as attractive as possible to potential purchasers of shares.[54] Because of the absence of effective competition, it has been necessary to create bodies such as Oftel and Ofgas to supervise the privatised utilities. The complexity of a privatised electricity industry will be such that the regulator will have a particularly crucial role in determining the commercial relationships that develop between generators, distributors and customers.

The growth of regulation is not confined to privatised public utilities. A new system of regulation has been introduced for the financial services industry. Although it incorporates a considerable element of self-regulation, its impact is evident in the advertisements which frequently appear in the financial pages of the newspapers from firms seeking compliance officers to police the regulations within their companies. Moreover, the system of regulation is likely to be tightened further in the future. Perhaps the most important area of increased regulatory activity is that of environmental and health and safety regulation, with particular implications for industries involved in potentially high risk processes such as the chemical industry. This is an area in which the European Community is increasingly active, leading to far-reaching changes in the scope, and methods of implementation, of British regulations.

These increases in regulatory activity have taken place against the background of a government effort to encourage deregulation, a policy initiative which has run into obstacles arising from the prevalent Whitehall culture, European Community requirements, and public demands for increased consumer protection.[55] The broader significance of these changes in terms of the concerns of this chapter is they make it even more difficult to assess whether reduced government involvement in the economy has had beneficial consequences, because intervention has diminished in some respects but increased in others.

The Thatcher government claims that 'Since 1981 the economy has been transformed.'[56] 1981 is chosen as a base date because the recession of 1979–81 is seen as demonstrating 'all too clearly the weakness of an economy which, over many years, has been uncompetitive and unresponsive to change.'[57] As it so happens, choosing a 1981 start date for measures of industrial performance does give a more favourable picture than the use of a 1979 base.

Assessing the impact of the Thatcher government's policies on manufacturing industry is not possible within the scope of this chapter. It would, for example, require an extensive discussion of exchange rate policy. Moreover, much discussion of the subject often proceeds from two irreconcilable positions. One viewpoint regards Mrs Thatcher as an inspired political leader who has revolutionised an outdated, inefficient economy labouring under the burden of overpowerful unions and excessive government intervention. The opposite position sees her as having fundamentally weakened Britain's industrial base, in the process increasing divisions between the north and south of the country.

Any assessment is ultimately going to depend on value judgements. For example, is a given share of manufacturing in GDP and employment a desirable and important objective? Should a chronic and substantial balance of payments deficit be a major concern? For example, the 1985 report from the House of Lords Select Committee on Overseas Trade, chaired by Lord Aldington, drew on evidence from senior industrialists to express grave concern about the performance of the British industrial economy. It was, however, dismissed by its critics as 'including almost every cliché and commonplace everyone has uttered about the British economy, and tries to shift the meaning of these clichés by substituting "manufacturing" when it ought to say "output" and national economy.'[58]

It is arguable that the government's distate for interventionist policies has sometimes led them to abolish institutions and discontinue policies which were making a positive contribution to competitiveness. Nowhere has this been more evident in two continuing areas of deficiency in British performance, training and research and development.[59] Although the old statutory training boards (now all to go apart from that for construction) had their deficiencies, the non-statutory training organisations (in effect, employers' associations) which have replaced them have not, in general, made an effective contribution to reducing Britain's deficiencies in skilled manpower. In the area of research and development, the government has withdrawn from 'near market' research and development. In 1988 it terminated the group of assistance programmes marketed under the 'Support for Innovation' label, including such useful schemes as the Flexible Manufacturing Systems scheme for computer-controlled batch production and the Fibre Optics and Opto-Electronics Industries Scheme. In this case, ideological preferences can be claimed to have overridden competitiveness objectives.

5 Lessons from the past and lessons for the future

Any government wanting to intervene systematically to improve the competitiveness of British industry would need a rather different bureaucratic apparatus from that provided, for all its strengths, by the British civil service. In the case of the steel industry, where government had a larger number of civil servants at its disposal than is usual for the conduct of relations with a particular industry, 'although the quality of BSC decision-making left much to be desired, both Ministers and officials had only rarely considered themselves sufficiently qualified to challenge the authority of the Corporation.'[60] If a future government decided that it needed expertise within the bureaucracy on particular industries (as most other western governments, including that of the United States, do find helpful), it might be advisable to reconstitute industry divisions so that they drew on a mixture of individuals with experience of the industry concerned, economists and other academic experts, and generalist civil servants.

Although it is difficult to draw out from the influential literature on flexible specialisation any clear recommendations on how and where government should intervene to support the efforts of entrepreneurs, it would seem that there is much to be said for public-private partnerships which combine government funding with relevant private sector expertise. One of the more successful experiments in intervention in Britain has been the Scottish Development Agency which has operated outside the civil service and which has recruited extensively from the private sector.[61] There has been some resentment from Conservatives that 'the SDA often gets public credit which they feel should go to the [Thatcher] Government.'[62] It is planned that by 1991 it will be recast as Scottish Enterprise, taking over the functions of the Training Agency in lowland Scotland, and operating through twelve local enterprise companies whose boards will have a two-thirds private sector membership. Some scepticism has been expressed about these changes, but the development agency model could usefully be applied in other parts of Britain apart from Scotland, Wales and Northern Ireland. Indeed, it has been developed at the local level through a variety of enterprise agencies.[63]

6 Conclusions

'The path of the British economy since the late nineteenth century has been one of persistent decline relative to its main industrial competitors.'[64] As has been pointed out in this chapter, the policies of successive British governments to improve manufacturing competitiveness have been characterised by hesitancy and uncertainty, and such interventions as have taken place have generally been *ad hoc*, sporadic and incoherent. An underlying reason has been a profound uncertainty, which has never been resolved, about what government's relationship to industry should be. The relationship with industry has been made more difficult by the separation of industrial and bureaucratic careers, and by the character of a civil service in which stability, smoothness of operation and avoiding embarrassment for ministers often seems to be the first priority of government.

Even so, as the chapter by Brown on agriculture shows, government has been able to intervene in that policy area in a way which has had an impact on the structure of the industry (much the same could be said of social policy areas such as health policy). As Brown's chapter also points out, the consequences in terms of competitiveness have been less substantial than is sometimes claimed, quite apart from the impact on the countryside. Policy which is effective in the limited sense that it has an impact would seem to require a relatively sheltered policy arena in which agreement about the principal objectives of policy can persist over a long period of time, as has been the case in agriculture.

Those who call for a more consistent industrial policy often overlook the fact that such a policy might have further slowed the process of industrial adjustment in Britain. Successful adjustment in the future will, however, depend on the presence of a skilled and flexible workforce and increased levels of investment in research and development. These tasks cannot be left to market forces, but do require government action to provide resources, to impel employers to pay attention to longer-term needs, and to ensure that the institutional arrangements for policy implementation are well designed.

Notes

1 National Economic Development Office, *British Industrial Performance*, London, 1984, not paginated.

2 B. W. E. Alford, *British Economic Performance 1945–1975*, London, 1988, p. 19.

3 A. Cairncross, 'What is de-industrialisation?', in F. Blackaby (ed.), *De-Industrialisation*, London, 1979, p. 10.

4 *Financial Times*, 12 October 1989.

5 *Ibid.*

6 See N. Harris, *Competition and the Corporate Society*, London, 1972.

7 *The Right Approach: a Statement of Conservative Aims*, p. 18.

8 *Financial Times*, 17 November 1989.

9 *OECD Science and Technology Indicators Report No. 3 — R & D, Production and Diffusion of Technology*, Paris, 1989.

10 G. Maynard, *The Economy under Mrs Thatcher*, Oxford, 1988, pp. 164–70.

11 See M. J. Piore and C. F. Sabel, *The Second Industrial Divide*, New York, 1984.

12 A. Daly, D. Hitchens and M. Wagner, 'Productivity, machinery and skills in a sample of British and German manufacturing plants', *National Institute Economic Review*, February 1985, pp. 48–61 and H. Steedman and K. Wagner, 'A second look at productivity, machinery and skills in Britain and Germany', *National Institute Economic Review*, November 1987, pp. 84–94.

13 E. E. Williams, *Made in Germany*, London, 1896.

14 See D. Finegold and D. Soskice, 'The failure of training in Britain: analysis and prescription', *Oxford Review of Economic Policy*, 1988, 3, pp. 21–53.

15 Cm. 278, *DTI — the department for Enterprise*, London, 1988, p. 1.

16 K. Middlemas, *Politics in Industrial Society*, London, 1979, p. 374.

17 *Ibid.*, p. 380.

18 See W. Grant, J. Nekkers and F. van Waarden (eds), *Organizing Business for War*, Oxford, forthcoming.

19 C. Johnson, *MITI and the Japanese Miracle,* Stanford, 1982, pp. 63–73.

20 There is a three-week course at the Civil Service College on government and industry for administrative trainees which includes placements with companies.

21 W. Lazonick, 'The cotton industry', in B. Elbaum and W. Lazonick (eds), *The Decline of the British Economy*, Oxford, 1986, p. 21. See also J. H. Bamberg, 'The rationalisation of the British cotton industry in the interwar years', *Textile History,* 1988, pp. 83–102.

22 S. Tolliday, 'High tide and after: Coventry's engineering workers and shopfloor bargaining, 1945–80', in B. Lancaster and T. Mason (eds), *Life and Labour in a Twentieth Century City: the Experience of Coventry*, Coventry, undated, pp. 208–15.

23 *Ibid.*, p. 216.

24 P. Hall, *Governing the Economy*, Cambridge, 1986, p. 66.

25 R. P. T. Davenport-Hines, *Dudley Docker*, Cambridge, 1984, p. 114.

26 J. Turner, 'The politics of business', in J. Turner (ed.), *Businessmen and Politics*, London, 1984, p. 13.

27 *Ibid.*, p. 13.

28 For the formation and subsequent history of ICI, see W. J. Reader, *Imperial Chemical Industries: a History*, Oxford, 1975.

29 C. Kennedy, *ICI: the Company that Changed Our Lives*, London, 1986, p. 22.

30 W. Grant, W. Paterson and C. Whitston, *Government and the Chemical Industry*, Oxford, 1988, p. 6.

31 See L. Hannah, *Electricity Before Nationalisation*, London, 1979.

32 On milk marketing, see S. Baker, *Milk to Market*, London, 1973. On the formation of the British Sugar Corporation in 1936, see Sir J. Winnifrith, *The Ministry of Agriculture, Fisheries and Food*, London, 1962, pp. 71–5.

33 R. S. Sayers, *The Bank of England, 1891–1944, Volume 2*, Cambridge, 1976, p. 547.

34 G. C. Allen, *The British Disease*, London, 1979, p. 51.

35 M. W. Kirby, 'Industrial policy', in S. Glynn and A. Booth (eds), *The Road to Full Employment*, London, 1987, pp. 134–5.

36 Tolliday, 'Tariffs and steel, 1916–34: the politics of industrial decline', in J. Turner (ed.), *Businessmen and Politics*, London, 1984, p. 74.

37 S. Tolliday, *Business, Banking and Politics*, Cambridge, Mass., 1987, p. 336.

38 *Ibid.*, p. 287.

39 *Ibid.*, p. 292.

40 J. Turner, 'The politics of business', p. 3.

41 M. W. Kirby, 'Industrial policy', p. 138.

42 C. Barnett, *The Audit of War*, London, 1986.

43 The results of the work of the Anglo-American Productivity Councils are summarised in G. Hutton, *We Too Can Prosper*, London, 1953.

44 A. D. Morgan, 'Commercial policy', in F. Blackaby (ed.), *British Economic Policy 1960–74*, London, 1978, p. 555.

45 See J. Leruez, *Economic Planning and Politics in Britain*, Oxford, 1975, pp. 87–9.

46 S. Wilks, *Industrial Policy and the Motor Industry*, Manchester, 1984, p. 118.

47 J. J. Richardson and G. F. Dudley, 'Steel policy in the UK: the politics of industrial decline', in J. Mény and V. Wright (eds), *The Politics of Steel*, Berlin, 1987, p. 334.

48 J. Mény and V. Wright, 'Introduction: state and steel in Western Europe', in J. Mény and V. Wright (eds), *The Politics of Steel*, p. 21.

49 *Financial Times*, 9 November 1988.

50 See J. Lambert, *Government Economic Service Working Paper No. 61*, 'Clothing industry scheme', London, 1983.

51 For a discussion of the relevant evidence, see W. Grant, 'Large firms and public policy in Britain', *Journal of Public Policy*, IV, pp. 1–17.

52 Cm. 278, p. iii.

53 Commission of the European Communities, *First Survey on State Aids in the European Community*, p. 13.

54 See M. Bishop and J. Kay, *Does Privatisation Work?*, London, 1988.

55 See G. Ashmore, 'Government and business: reducing red tape', *Public Money and Management*, VIII, 1 & 2, pp. 78–81.

56 Cm. 278, p. 2.

57 *Ibid.*, p. 2.

58 Samuel Brittan, 'Coronets and begging bowls', *Financial Times*, 17 October 1985.

59 See S. Prais and K. Wagner, *Schooling Standards in Britain and Germany*, NIESR Discussion Paper 60, London, 1983; E. Keep and E. Mayhew, 'The assessment: education, training and economic performance', *Oxford Review of Economic Policy*, 1988, III, pp. i–xv; P. Stoneman and J. Vickers, 'The economics of technology policy', *Oxford Review of Economic Policy*, 1988, IV, pp. i–xvi.

60 J. J. Richardson and G. F. Dudley, 'Steel policy in the UK', p. 354.

61 See W. Grant, *Government and Industry*, Aldershot, 1989, pp. 106–10.

62 *Financial Times*, 25 October 1988.

63 For an overview, see I. Barnes, J. Campbell and J. Preston, 'Is local authority economic intervention a waste of resources?', *Public Policy and Administration*, 1987. pp. 1–8.

64 K. Smith, *The British Economic Crisis*, Harmondsworth, 1984, p. 55.

7 *Geoffrey Jones*

Competition and competitiveness in British banking, 1918–71

1 Introduction

The inclusion of a chapter on domestic banking in a volume concerned with the impact of government policy on Britain's international competitiveness may seem, at first sight, curious. Yet there has been a long tradition of criticism of the banks for their allegedly adverse effect on British economic performance in the twentieth century. It has been maintained, for example, that the banks 'failed' British industry by credit allocation and other policies which were inferior to those followed by German, American and Japanese banks. A more recent line of criticism is that the British banks failed to act as 'visible hands' to discipline and reconstruct inefficient industries. Moreover, there have been allegations that government macro-economic policy was unduly weighted towards serving the interests of the banking sector or the 'City'.[1] The banks figure highly as important, if shadowy, players in the story of Britain's declining competitiveness.

This chapter, which draws on the series of fine studies on British banking which have recently appeared,[2] begins by looking at the issues of competition and competitiveness *within* the banking sector between 1918 and 1971. Theory provides no consensus how these two things are linked. While some writers have regarded vigorous internal competition as an important stimulus to competitiveness, other traditions have seen restricted competition as providing a secure base for innovation. Section 3 turns to the central issue of this volume, the role of the government. Throughout the twentieth century governments have been closely involved in the banking industry in Britain and elsewhere, both as borrowers and regulators. Again, however, theory provides conflicting views on the likely outcome of such government activity. While some would see government intervention as essential given the asymmetry of information in many financial markets, supporters of 'free banking' such as Hayek blame state interventionism for excessive conservation and lack of innovation, alongside many other difficulties.[3] In Section 4, the implications of the competitive situation in banking on competitiveness elsewhere in the British economy is considered.

2 Competition and competitiveness

By 1918 the oligopolistic structure of British domestic banking was already in place. Largely through mergers and amalgamation, the considerable number of banks in England and Wales in the mid-nineteenth century had been reduced to a small number of large banks by 1918. Scottish banks operated under separate legislation, but a similar concentration occurred in the 1920s and during that decade many formal links were created between the banks in the two systems. By 1920 the largest five banks in England and Wales held 80 per cent of total bank deposits. Over the following fifty years there were amalgamations of smaller banks, but no great structural change until 1968 when two of the 'Big Five' merged to form National Westminster. By this date retail deposit banking in Scotland was dominated by three banks, Clydesdale, Bank of Scotland and Royal Bank of Scotland (all with English 'Big Four' shareholders).[4]

Within this structure, price competition was extremely limited. Price matching is to be expected in an industry with a homogeneous product, but there was also extensive collusion with long-standing cartel arrangements. At the heart of the cartel in England and Wales was the London Clearing House, an arrangement which dated back to the eighteenth century. Membership was restricted: if a bank was not a member it was difficult to make much progress in retail banking.[5] New members were allowed, but membership was granted only sparsely. In 1936, for example, the small District Bank was allowed to join. The London Clearing House provided a means through which the collusive agreements on prices were reached. At least from the mid-nineteenth century, agreements on rates were becoming common. After 1920 regular meetings of the London Clearing Bankers Committee fixed deposit and advance rates.[6] In Scotland, which had its own clearing house, there was widespread collusion on interest rates from at least the 1850s through a series of 'Agreements and Understandings' between the banks. These grew in scope and stability over time.[7]

The high level of concentration and price collusion did not, however, exclude non-price competition. Recent advances in the theory of contestable markets have demonstrated that the degree of competition cannot be neatly equated with the number of firms in an industry.[8] Freedom of entry into banking was a key factor in the long term, which will be discussed below, but it is also important to note that competition of some kinds between the clearing banks never disappeared.

In the decades after 1918 the large British banks competed with one another for deposits and, even more, for corporate accounts. In the 1920s, this competition seems to have been vigorous. Lloyds opened many new branches in that decade, and even allowed managers in the regions to vary their rates 'to meet local competition'.[9] Pre-credit card retail banking encouraged high branch density. Midland Bank opened 638 new branches in the 1920s (to give a total of 2,044 by 1929) and introduced a variety of new products.[10] The Scottish banks also opened many new branches and competed through improvements in quality of their services.[11] The bankers who testified to the MacMillan Committee on Finance and Industry in 1930 suggested competitive forces were fiercer than in the days before the great merger wave. The Chairman of the National Provincial Bank considered that 'the competition is quite disgraceful at the present moment', while Lloyds' Chairman contrasted the situation in 1930 with his experience as a private banker in the North of England 'in the old days':

> Formerly we were almost exclusively banked by private banks in the North; the partners of the different firms used to dine at each other's houses, they were on friendly terms and there was a feeling that it was not quite fair to take away accounts. I think competition was certainly less in those days than it is now.[12]

Studies of the inter-war British steel, motor car and cotton industries testify to the continued competitive forces in English banking in this period. In the steel industry, banks avoided interference in the affairs of even unprofitable clients for fear that they would take their business elsewhere. A similar phenomenon existed in the motor industry.[13] The Lancashire cotton industry in the 1920s could draw on five 'local' banks as well as branches from the Big Five, especially Midland and Barclays. All these banks were caught with large frozen advances after the collapse of the 1919–20 boom. The competitive nature of Lancashire banking in the 1920s was a major obstacle to structural change in the cotton spinning industry. 'In the competitive structure of the Lancashire banking', one study has concluded, 'any bank which enforced liquidations or amalgamations of mill companies was likely to suffer a loss of customers to its rivals, who would be able to cream off the best accounts, leaving the enforcing bank with a high concentration of doubtful debts'.[14] As a result of the over-exposure of the Lancashire banks to the cotton industry and of the cotton industry's need for rationalisation, Montagu Norman, the inter-war Governor of the Bank of England, devoted much time and energy trying to encourage a merger of those banks into a single institution — and thereby create a 'Big Six' in English banking — but although amalgamations occurred, his grand scheme was never realised.[15]

The available evidence suggests a general decline in non-price competition between the banks in the 1930s. Both the experience of banks with troubled industrial clients and encouragement from the Bank of England prompted more co-operation. The Bank of England led the clearers into a series of schemes designed to facilitate the rationalisation of the basic industries, most notably the Bankers Industrial Development Company. The banks joined together in rescues of failed businesses, such as the Royal Mail group of shipping companies in the early 1930s, and generally there was an increased 'level of cooperation between the major banks in the running of large industrial and shipping accounts'.[16] In the 1930s, with cheap money, industrial firms met their financial needs by selling government securities. The demand for bank advances fell as a result, reducing the incentive for competitive bidding for accounts. The banks responded by transferring their funds from advances and investing instead in gilts. The ratio of investments to deposits of London clearing banks rose from 14.4 per cent to 28 per cent between 1929 and 1937, while the ratio of advances to deposits declined from 52 per cent to 41 per cent over the same period.[17]

Although competition for good corporate accounts never ceased, it is unlikely that the 1950s saw much improvement over the 1930s in the degree of competition on British domestic banking. Government controls on bank lending were, as discussed below, a big disincentive to competition. Investment ratios remained high until 1958, with the government's financial requirements rather than those of the private sector dominating the asset portfolios of the clearing banks. In the late 1940s, Midland Bank had 75 per cent of its deposits lent to the government in one form or another.[18] Interbank agreements extended further into non-price competition. The clearers agreed not to take accounts from one another when advances were refused 'as being contrary to the criteria drawn up by the authorities', and various 'self denying ordinances' were reached under which the clearing banks agreed not to compete on interest rates or commission charges.[19]

It took the three years' suspension of direct government credit controls and guidelines between 1958 and 1961 before there was much reawakening of competitive vigour among the clearing banks. This was reflected in the balance sheets of the London clearing banks which showed a dramatic rise in advances between 1958 and 1960 (and resulting fall in investments), followed by a more modest rise in advances over the 1960s.[20] There was another surge in branch opening, with the total number of bank branches in England and Wales expanding by over 1,000 (to reach 11,804) between 1958 and 1964. The 1960s also saw more product innovation, with the clearing banks showing a greater interest in medium-term finance and diversifying into new fields, such as hire purchase.[21] A Prices and Incomes Board investigation in 1967 noted that, although price competition was restricted, the banks 'compete strongly with each other, imposing costs upon one another through the competitive duplication of branches'.[22]

There was, therefore, never a period between 1920 and 1970 when the clearing banks stopped competing with one another, but at times the competitive light flickered and it was generally confined to matters other than price.

This was not, however, the end of the story, for British banking was never wholly synonymous with the clearing banks. The Clearing House system made entry into the cartel of the clearers extremely difficult — but entering into competition with the cartel was relatively simple. No regulations prevented those outside the cartel taking deposits, nor did the government regulate interest rates for all institutions. Nor were foreign institutions prevented from establishing branches in Britain if they met basic prudential requirements.[23] As a result, while the twentieth century saw restricted competition between the clearing banks, it also saw a strong competitive assault on their position. As a result, their market share fell sharply over time, suggesting — for one reason or another — a formidable long-term decline in competitiveness.

This is the view taken by Michael Collins, who dates the declining market share of the clearers at least to the early twentieth century, and suggests that 'whereas the root cause lay in the long-term process of financial sophistication, the banks' competitive stance was an important contributory factor'.[24] During the inter-war years the banks lost market share of deposits to other institutions, notably the building societies, the Post Office Savings Bank and the trustee savings banks. Their relative decline continued at a faster rate during the thirty years after 1945. The charts in the Appendix illustrate the clearing banks' declining market share of deposits. Even more noteworthy was the decline of market share in lending. By 1980 foreign banks were providing almost a third of all bank loans to British manufacturing industry.[25] In 1957 the 16 London and Scottish clearing banks held around 86 per cent of the total identified assets of banks in the UK. By 1978 their proportion of total assets of banks in the UK was down to 23 per cent, and even their proportion of sterling assets was only 48 per cent.[26] It is never easy to make neat comparisons between banking and manufacturing, but there would appear to be a case for comparing the declining competitiveness of the British clearing banks with the difficulties of, say, the British-owned part of the motor industry from the 1950s.

The two winners in the competitive struggle against the clearing banks were the building societies and foreign banks. Although the British fiscal system played a role in diverting savings towards building societies, recent studies of their strategies in the inter-war years have stressed 'the vigorous institutional response of building societies to their market and regulatory environment'.[27] They experimented and innovated new products and ideas. After the war they again proved efficient, well-managed institutions which competed for deposits by advertising, branch expansion and quality of service. The total number of building society branches grew from just over 1,500 in 1955 to 2,500 in 1970, and 6,000 by 1980. The total number of building societies was 480 in 1970, falling steadily to 270 in 1980.

Foreign banks had established branches in London since the late nineteenth century, but it was with the emergence of the Eurocurrency markets at the end of the 1950s that they began to appear in substantial numbers. By 1975 some 35 American banks had branches in Britain. Initially the

American banks largely participated in the Eurocurrency markets. Over time they began to make loans to American and other foreign-owned customers in Britain, largely for international trade finance. By the late 1960s, and especially after 1971, they began to develop a substantial business in sterling lending to British customers, mostly companies, a business largely financed from the inter-bank market rather than retail deposits. As with the building societies, among the factors encouraging the relative growth of the American banks was product innovation, notably term lending. In 1977 almost 30 per cent of the American banks' sterling loans and 70 per cent of their loans in other currencies had an initial maturity exceeding one year. They also employed the kind of aggressive marketing techniques to corporate customers which the British clearers were to adopt only in the 1970s, if not later.[28]

3 The government, the Bank of England and the banks

To what extent did the apparent decline in competitiveness of the clearing banks, and the restricted nature of competition between them, stem from government policies?

In a comparative perspective, and especially in contrast to the United States, Britain had no developed system of external supervisory regulation of banks until the lessons from the secondary banking crisis of 1973–74 and EC directives pushed Britain along that road. Before the Banking Supervision Act of 1979, British banking essentially was 'self-regulating'.[29] British bankers lobbied hard in the inter-war years and afterwards to resist direct government regulation of their market activities.[30]

The complication here is the role of the Bank of England. In terms of ownership, the Bank of England was not a part of the government until its nationalisation in 1946. Yet the Bank's slow transition over the nineteenth century (despite the provisions of the Bank Charter Act of 1844) into a non-competing and non-profit maximising central bank which undertook the monetary management of the British economy made it in effect an agent of the government. The Bank of England certainly represented City interests to the government, but it is not clear if it was any more or less 'captured' by its constituents than a state-owned regulatory agency. By the inter-war years, the Treasury and the Bank of England worked together on banking matters, and it surely makes sense to treat Bank policies towards the banks as proxies for government ones in this period. Nationalisation made little difference to this situation, and until the 1970s the Bank remained the main agency in government dealings with the banks. As the clearers told the Wilson Committee in 1977:

> the banks' attitude had long been that lines of communication between themselves and Whitehall and Westminster should run via the Bank of England. The Governor

of the Bank was seen as acting both as the government's representative in the City and as the City's ambassador to Whitehall ... they were very firmly wedded to the view that they should have little contact with government other than their traditional relations with the Bank of England.[31]

The quasi-autonomous position of the Bank of England, under both private and public ownership, has been regularly commented upon by critics of government economic policy-making. Peter Hall (among others) has seen the Bank as a 'powerful force for fiscal conservatism' because of its concern for the international position of the City, and an obstacle to the adoption of 'a more dirigiste industrial policy'.[32] It might be as fruitful to examine the relationship between the Bank and the Treasury in terms of principal-agent theory. In the 1950s there was serious conflict between the two parties on Bank Rate, for example, and it would not be surprising if efficiency gains could not have been achieved by a more appropriately specified contract.

For present purposes, however, the key point is that the Bank of England, as the representative of the monetary authorities, exercised a considerably greater influence on the banking system than the lack of formal regulation would suggest. By the inter-war years the Bank of England and the clearers coexisted in a harmonious relationship. The Bank made its wishes known through nods and winks rather than regulatory directives. The banks could in theory not accede to the Governor's wishes — and Barclays and Lloyds ignored them for two decades over the matter of their overseas banking interests — but the era of turbulent clearing bank chairmen (such as Edward Holden of the Midland Bank) was over, and a radical refusal to follow the Governor's wishes on a major matter never occurred. The Bank of England's formal relations with the banks passed through the Committee of London Clearing Bankers.[33] The Bank, as a participant in the markets as well as a regulator, almost certainly had more information about 'its' industry than any British government department in the inter-war years, or indeed much later.

In its policies towards the banks, the primary Bank of England concerns were the 'public interest' (defined by a dialogue with the government) and the security of depositors. The Bank was very much less concerned with either the shareholders of individual banks or bank borrowers. 'Public interest' concerns led the Bank to welcome the concentration and cartelisation of the clearers. These developments facilitated the Bank's control over monetary policy from the nineteenth century. This control depended to a large extent on the willingness of the banks to follow the Bank of England's lead, without coercion, and it was far easier to exercise such an influence over a small number of institutions. The Bank sought to influence interest rates and the allocation of credit within the economy by, among other devices, varying its discount rate — the Bank Rate. Concentration in British banking and the rigidity in the interest rate structure simplified the task of making the Bank Rate effective by ensuring that the market rate on bills was close to the Bank Rate. The Bank of England — and the government — came to put enormous value on their ability to control monetary policy through the banks. When the clearers sought to dismantle their cartel in the early 1960s they were blocked, and in 1969 the government again told them that any abandonment of agreements on deposit and lending rates would be against the public interest.[34]

A concentrated and cartelised banking structure also fitted well with another priority of the Bank of England, banking stability and — ultimately — the protection of depositors. The Bank welcomed and sometimes orchestrated collective moves by the clearers in the interests of stability. During the inter-war years, as already observed, the Bank encouraged the collaboration of the clearers in a number of industrial reorganisation schemes. When Britain's largest overseas bank, the Anglo-South American Bank, virtually collapsed in 1931 the Bank orchestrated a 'pool' of the clearing banks to come to its rescue.[35] In the mid-1930s the Bank, anxious to guarantee a full subscription of Treasury bills, arranged a cartel among the discount houses, which had been badly affected by the recession, to bid at agreed rates for Treasury bills. The clearers, who also had competed for Treasury bills in the 1920s, agreed not to tender for bills and to lend to the discount houses on call at low rates.[36] These arrangements remained in force until 1980. The advances lock-ups in the 1920s, and then the 1931 financial crisis, gave powerful levers to the Bank of England. And the Bank wanted stability and not adventure, not least for concern that 1931, from which the British banking system escaped lightly compared to Germany and the United States, must not be allowed to recur.

Apart from encouraging collaboration and the avoidance of price competition, the Bank of England — supported by the Treasury — froze the industry structure after the end of the First World War. Concern at the pace of bank amalgamations led to the Colwyn Committee, which in its Report in 1918 recommended legislation requiring prior approval of any further amalgamations. No legislation ever materialised, but it was established policy that the Big Five in England and Wales would refer any amalgamation among themselves to the Treasury for approval, and a set of rules to this effect was codified in 1924.[37] It was only at the end of the 1960s that this position was materially changed, when an indication from the government that a merger among the Big Five might be allowed led to the amalgamation of Westminster and National Provincial, although another proposed merger between Barclays and Lloyds was disallowed.[38] While forbidding mergers, the Bank of England effectively made the Big Five bid-proof. A foreign take-over of a British clearing bank was not allowed until 1987, when National Australia Bank was allowed to purchase Clydesdale Bank, the smallest of the Scottish clearers, from the Midland Bank.[39] It was not until 1990 that a foreign purchase of a small English domestic bank was permitted, when National Australia Bank purchased Yorkshire Bank from the group of British clearers which owned it. Market assessment of the performance of the clearers was obstructed because bankers were exempted under the Companies Acts from declaring their 'true' profits, and it was not until 1969 that the London clearing banks agreed to full disclosure.

The highly specialised nature of the British financial system was encouraged and protected by the Bank of England and the Treasury. Each part of the banking system had its own association — the Accepting Houses Committee for the merchant banks, the London Discount Market Association for the discount houses, the London and Scottish Clearing Bankers for the clearers — through which the Bank of England conducted its formal business. The government operated different regulatory authorities for different parts of the financial system. While the Bank of England supervised banks, the Board of Trade regulated hire purchase companies and the Chief Registrar of Friendly Societies supervised building societies. The Bank of England took a conservative view of what the 'proper' role of each part of the financial system was and discouraged innovation, especially by the clearing banks. During the 1950s, for example, the plans of several Scottish and English clearing banks to enter the hire purchase business were blocked for some years by the Bank of England view that 'hire purchase was not the proper business of a bank'.[40] There is clear evidence that, at least during the 1950s, entrepreneurial responses within the banks to their declining market share were being obstructed by the authorities.

Government controls on bank lending proved a highly effective dampener on competition and innovation. From the end of the Second World War until 1971 (and indeed later) lending by the clearing banks was subject to official controls in the interests of overall government macro-economic and monetary policy. The allegedly persistent weakness of the balance of payments justified a position whereby the banks were expected to be handmaids of official policy towards sterling. There were both qualitative and quantitative controls. The banks were asked to give priority to certain 'strategic' sectors, notably defence, exports and import saving activities. In 1955, in the middle of a balance of payments crisis, an actual reduction in bank advances was called for, and the London clearing banks aimed at a 10 per cent reduction. Further quantitative lending controls were introduced in 1957. Between 1958 and 1961 controls were lifted — with the result, as already noted, of a noteworthy increase in bank lending — but in the 1960s restrictions again came to the fore. From 1960 the London and Scottish clearing banks were obliged to make 'special deposits' with the Bank of England. Between 1965 and 1971 there were official 'ceilings' for loans. A limit on the growth of clearing bank lending to the domestic private sector of 5 per cent per annum was applied in 1965, of zero at the end of 1967, and of 4 per cent in May 1968. In November 1968 there was a reduction of 2 per cent over four months (except for export and shipbuilding finance) followed by further limits on lending growth until 1971 when the system was abandoned.

Recent research on the effectiveness of these controls over bank lending in the 1950s suggests that bank lending in this period did move towards the sectors favoured by the government, although the various controls had less of a discernable impact on the actual levels of internal business activity and the pressure of demand.[41] Competitive instincts within the banks were certainly dampened. Far from competing for new business, by the 1950s branch managers of the Clydesdale Bank were being sent congratulations by their seniors for reducing their lending, in accordance with Treasury rules. Aggressive pursuit of deposits was hardly encouraged given the restrictions and limitations on how they could be lent.[42]

Post-war government controls on lending, therefore, restricted and discouraged competition between the clearing banks. They also worked to undermine the competitiveness of the sector as a whole by being applied in a discriminatory fashion against the clearers.[43] The clearing banks' greatest competitors for retail deposits, the building societies, were assisted by a special system of taxing the interest on building society deposits — the composite rate of tax — which gave them some advantage over their banking competitors and enabled them to offer a marginally higher rate of interest. There were no restrictions on the level of loans by the building societies, beyond prudential controls, which enabled them to expand at a fast rate, and gave them an incentive and reason actively to pursue deposits. The quantitative lending controls of the 1950s were only applied to the clearing banks. They, too, were alone in being obliged to maintain a ratio of liquid assets to deposits of 30 per cent of which cash had to make up 8 per cent. In addition, the special deposit scheme in operation after 1960 applied only to the London and Scottish clearing banks. When credit ceilings were reintroduced in 1965, they were finally applied to non-clearing banks and finance houses, but the other institutions were still not obliged to meet the liquidity ratios set for the clearers or make special deposits. 'The official restrictions', the Committee of London Clearing Bankers observed in 1977 of the pre-1971 era, 'contributed greatly to the loss of a large part of the clearing banks' share in total lending over the period.'[44]

The non-clearing bank financial intermediaries did not, however, escape all official policies. The Bank of England was prepared to cast its net more widely than the clearers in pursuit of its goals of stability and the discouraging of product innovation. In the early 1950s, for example, the Treasury and Bank of England moved to curb the finance of hostile take-over bids by insurance companies and foreign banks. Such bids were a new phenomenon and were regarded both as 'speculative' and likely to lead to unwanted credit expansion. In December 1951, the Bank of England issued a 'request' to the clearing banks not to finance 'speculation' in securities, and two years later this 'request' was extended to the insurance companies

— which had played a significant role in financing several early take-over bids through the purchase and leaseback of buildings — and American banks. The American banks in London responded to the Governor of the Bank of England assuring him that they were 'glad to co-operate and to be guided by the wishes which had been expressed'.[45]

It does not require great flights of fantasy to envisage alternative regulatory, fiscal and credit policies which would have encouraged more competition between the clearing banks, and not worked so single-mindedly to undermine their competitiveness. It is another matter whether such alternatives were realistic given the historical legacy of British banking and the fact that, after 1945, policies towards bank lending were part and parcel of the overall British government preoccupation with inflation and the balance of payments. Curiously, however, a vision of a counterfactual world was in place in the international banking activities of the City of London. Rumours of London's demise as the world's leading international financial centre in favour of New York after the First World War have been exaggerated, but in any event it seems clear that whatever happened in the inter-war years, London had re-established its prominence by the late 1950s, and has maintained that position ever since.[46] The emergence of the Eurocurrency market, based in London, played a vital role in re-establishing London's premier position. American controls over domestic interest rates on US bank deposits and over capital outflows made it more efficient to intermediate externally in London than in the United States. The Eurodollar market flourished in London not simply because of the existence of a large financial infrastructure, but because of the lightness of the regulatory controls. Although the Bank of England exercised a prudential watch over the Euromarkets, there were no liquidity ratios and there was freedom of entry and exit. As a result the markets came closer to 'free banking' than anywhere since the early nineteenth century. There was no attempt to interfere with market-driven innovation, and there was in fact an accelerating rate of innovation in financial instruments. The welter of regulations and controls over domestic British banks were not enforced in international banking. Foreign currency deposits and advances were exempted from the constraints imposed on sterling transactions in the 1960s.[47]

Under the alternative regulatory regime applied to international banking, London flourished from the late 1950s as the world's most competitive international financial centre. Although foreign, notably American, banks were major forces in the Eurocurrency markets, British clearing, merchant and overseas banks were also active participants in the fast growing and changing markets. If the regulatory conditions applied to international banking from the late 1950s had been applied to domestic banking, might not a more vigorous domestic banking system have ensued? The range of innovation and lending among the clearers between 1958 and 1961 during the brief period of relaxation of lending controls would appear to support such a view.

Two issues need to be resolved before a definite answer can be provided. The first is how far domestic banking markets can be equated with international banking markets. Information asymmetry is greater in domestic markets (which have large numbers of small depositors and lenders) which might suggest a greater need for regulation to enforce contracts. In addition, competition promotes efficiency in the long run by driving out weaker firms from an industry. However, in domestic markets this would put large numbers of depositors at risk, and it is unsurprising that governments and central banks have enforced regulations and erected barriers to exit to prevent banking failures.

A second consideration is whether the clearers have been capable of a vigorous and innovative response if regulatory conditions had been different. There have been many suggestions that, in Pressnell's words, 'competition on other than a restricted basis has never been popular with British banks'.[48] Agreements on rates certainly began to be popular well before the oligopolistic structure of the banking industry took shape, and by the 1950s many clearing bankers honestly regarded 'intensive competition' as being as much against the public interest as their own.[49]

Competitive instincts would have been improved by making the market for retail deposits fully 'contestable' by removing barriers to entry and exit, particularly the clearing house cartel.[50] However, there is also evidence that the process of amalgamation and merger had proceeded too far in the 1920s, and outrun the capacity of managers to run the large banks efficiently.[51] The solution lay either in an improvement in management methods (which in fact happened over time) or in the break-up of the large banks into more manageable units.

The clearers showed several signs of weak or inadequate management. In the inter-war years, when the banks ventured beyond the sheltered world of domestic banking to the Continent, they not only encountered defaults, depression and exchange controls, but also suffered from very evident managerial constraints.[52] After 1918 Lloyds accumulated, by acquisition, all the British overseas banks operating in Latin America, together with whole or part interests in banks and branch networks in India, Egypt and New Zealand, but right through until 1970 the opportunity was lost to mould these interests into a co-ordinated banking group, despite a series of pleas to that effect from within the Lloyds' management.[53] Midland has the most detailed and scholarly history of any English clearing bank to date, and that provides weighty evidence of management failure, especially for the 1930 to 1960 period. By the 1930s Midland had developed a highly centralised management structure to administer its large branch network, which appears to have contributed to the 'cautious and inward looking attitude' the bank developed. In the post-1945 decade Midland was afflicted by management 'inflexibility,' with branch managers 'stranded without the authority to lend even the smallest sum without reference to head office'.[54] Midland, in turn, restricted the competitive strategy of its Scottish subsidiary, Clydesdale. The same process was discernible in the other Scottish banks and this helps to explain the apparent contrast between the Scottish banks and other more vigorous parts of the Scottish financial scene, such as the insurance companies and investment trusts.[55]

Possibly there were problems with the staffing of banks after the First World War. The banks were loaded not only with many, not necessarily appropriate, staff acquired by mergers, but also with staff re-employed on coming back from the war. All this gave an imbalance of the age structure towards older, probably more conservatively attuned, men. Clearing banks in the inter-war years, and after, attracted men who wanted a 'safe job' and the result was, inevitably, a business culture more concerned with caution than innovation. Exogeneous circumstances — two wars, major recessions, the 1931 crisis, lending controls — would only have reinforced such sentiments.

The bank which seems to have devised the most effective management structure in this period was Barclays. After 1945, Barclays overtook Midland as Britain's largest clearing bank in terms of deposits. As Barclays lacks a scholarly history for this period, much remains unclear about the sources of its competitive advantage, but it is known that the bank had a very different management structure than Midland Bank. Modern Barclays Bank dated back to its incorporation as a joint stock bank in 1896 by the amalgamation of 20 private banks, and it retained a highly decentralised management structure from this inheritance. By the 1960s Barclays had a system of regional offices and local head offices. Each regional head office was headed by a regional general manager. Each local head office was headed by executive local directors, the larger ones having local boards including non-executive local directors. Barclays' decentralised structure of district head offices was identified by Midland's management (eventually) as a major factor in its rival's competitive success.[56]

If Barclays decentralised management structure produced competitive advantages — even within the cartelised restrictive environment of British banking — it might be suggested that the ideal solution for British banking would have been not by management decentralisation in existing groups but their break-up into fully independent banks. The existence of the highly profitable Yorkshire Bank may lend some credence to this view, notwithstanding that bank's serious problems in 1911, 1916 and in the early 1950s. The mergers of the late nineteenth century which created large units out of hundreds of small ones linked through the market by correspondent relationships may have carried the process of internalisation too far. Profit-maximising rationality may well have dictated a break-up of these large units but the barriers to exit and the ability of managements to restrict severely the information given to shareholders worked against such a process.

4 The banks and economic competitiveness

The main outlines of domestic banking in Britain between 1918 and 1971 are clear. It was a very concentrated industry with widespread collusion. The clearing banks were steadily losing market share. Government and Bank of England policies actively supported the collusion and, through their discriminatory impact, were one factor in the declining competitiveness of the clearing banks. However, the precise effect of this situation on other sectors of the economy, although much discussed, is unclear.

There were certainly costs in the British banking structure. The role of non-price competition in creating irrational preferences and supporting inefficiency is well known. Widespread duplication of branches was only the most visible aspect of such costs. The long-term stability of British banking was a major positive feature of the system, although this might have been achieved by alternative means. Brian Griffiths' assessment of the cost of the system to individual depositors carries weight:

> While it can be argued that the absence of interest rate competition among a few large banks meant that the risk of bank failure was very small and that the inefficiencies resulting from the cartel was a premium implicitly paid by bank depositors for the security of their balances, such a system is an extremely inadequate substitute for a system of deposit insurance and one in which the premium bears little relation to the risk.[57]

The system of regulation also imposed costs on the shareholders of the banks, notably through the liquidities ratios and special deposits of the 1950s and 1960s.[58]

As for the vexed question of the banks and the provision of finance for British industry, it is now common ground that twentieth-century British banks have always lent to industry, and that short-term overdraft facilities were often 'rolled over'. The British tradition was, in any case, for industrial finance to come from internal sources. As we have seen, the banks always competed for desirable corporate accounts. The structure of British banking affected not the flow of funds but the nature of the product on offer, for product innovation was visibly dampened by cartelisation and regulation. The banks were part of a highly specialised system, each part of which — clearing banks, merchant banks, discount houses, capital markets — appeared 'efficient', but what was really required was a greater flexibility in the definition of each part's role. A straight adoption of the Continental 'mixed banking' model, however, may not have been helpful. Although some British writers have praised this system, especially in Germany, before 1914, it has been less frequently observed that during the 1930s a series of

Continental countries — such as Belgium, Italy and Sweden — found mixed banking so unsatisfactory that they passed legislation against it, obliging banks to confine themselves to commercial business.

Because there was ease of entry (and exit) into British banking, as opposed to clearing banking, dissatisfied lenders and borrowers were able — and did — turn to other institutions than the clearing banks. During the 1950s merchant banks, insurance companies and other intermediaries began to discount large numbers of industrial bills in response to the restrictions imposed on the clearing banks. Yet the lack of innovation within what remained the largest holder of financial assets until the 1960s raised search costs for those seeking alternatives. Some of the alternatives were of more value to depositors than to industrial borrowers (notably the building societies). Others, such as the American banks, emerged as realistic alternatives only late in the period, following the development of wholesale money markets and after they had learned about the British market by lending first to the subsidiaries of American companies in Britain.

More innovation and price competition among the clearing banks would not have cured the basic problems of competitiveness in British manufacturing industries. Few researchers on the post-1945 period (or earlier) now believe that shortage or price of capital was a critical problem. The deficiencies in product quality which afflicted industrial competitiveness had their causes elsewhere, in education and in management training and structures. More dynamic banks might have provided a source of discipline to inefficient managements, a role which was much needed in the twentieth-century British economy, and which 'market forces' singularly failed to perform.[59] As it was, the banks at best refrained from interfering in the affairs of their clients: 'the bank's role as provider of finance and financial services must not be confused', as the banks put it in 1977, 'with the role of the company's shareholders and management'.[60] At worst, as the evidence from the inter-war steel, car and cotton industries suggests, they helped preserve inefficient managements and firms. It remains to be demonstrated, however, that British bankers had the ability and skills to improve the performance of such industries. If they had assumed a more active 'visible hand' role, they might have been less successful than the existing managements.

On a more macro-economic level, any claim that the banks and the City have helped to bias British policy over the long term towards their interests (and thereby damage competitiveness elsewhere) has to deal with the problem that, after 1945, restrictive and discriminatory controls contributed to a serious loss of market share by the clearers. Government policy may have supported London's role as a premier international financial centre, but it did not necessarily support the role of British-owned financial institutions. British economic policy between the 1940s and the 1970s, with its obsessions about sterling and the balance of payments, is open to criticism, but it would be a grave error to believe that these policies were the result of official pandering to the interests of the bankers, who were more victims that beneficiaries of official policies.

5 Conclusion

In 1971 the regulatory framework of British banking was radically altered by the Conservative Government's 'Competition and Credit Control' policy. New reserve asset ratios were introduced and applied uniformly to all banks and quantitative ceilings on lending were abolished. In return, the clearing banks abandoned almost all of their interest rate agreements. Although this represented a major policy change, a new era of unregulated competition did not happen overnight. In November 1973 the Heath Government, worried about a surge in money supply, introduced the supplementary special deposits scheme or 'corset'. The scheme was discontinued in 1975, reactivated in November 1976, and suspended once more in August 1977, reactivated and finally abolished in June 1980. It was not until the 'corset' was finally removed that the clearers felt it prudent to diversify into residental mortgages, a step they had wanted to take for a number of years.[61] Discrimination between financial institutions also continued. Between September 1973 and February 1975 the clearing banks were ordered to restrict the interest paid on deposits under £10,000 to a certain amount in order to help protect building societies from the effects of high interest rates. Throughout the 1970s the Bank of England continued to issue qualitative guidance on the direction of bank lending: in August 1977, for example, banks were told (as so often in the past) to give preference to exports and import-saving and to restrain lending to persons and property companies. Only in the 1980s was such guidance abandoned, and all forms of discrimination between the clearers and other financial institutions removed.[62]

Before the 1970s competitive forces within the clearing bank sector had been very restricted. Over time the clearing bank sector lost market share to other financial intermediaries. The clearing banks were in part responsible for their declining competitiveness. They benefited from their price cartel. Even within the contraints under which they operated, many clearing bankers took a narrow view of their function. There is evidence of management deficiencies. The process of concentration may have gone further than economic logic dictated, but there was no effective mechanism for disciplining inadequate managements. The clearers used the changed regulatory system after 1971 to slow the rate of decline in their market share,[63] but in the 1980s building societies and Japanese banks have waged further weighty attacks on their positions.

Whatever the failings of the clearing banks, it seems incontrovertible that the Bank of England and the Treasury were major influences in restricting competition between, and hampering the competitiveness of, the clearing banks, despite the lightness of external supervisory regulation. As in agriculture and oil, official policy placed a high priority on stability, and was not interested in competition. To facilitate control over monetary policy, the elimination of price competition was encouraged, approved and sustained. In pursuit of banking stability and depositor protection, collusive behaviour was permitted and promoted. The highly specialised British financial system was welcomed and the Bank of England discouraged and even forbade product innovation. After 1945 controls on bank lending further stifled competition and, by being applied in a discriminatory fashion, undermined the competitiveness of the clearing banks. The contrast between the regulatory regimes — and the consequences — in international and domestic banking could not have been greater. While, especially after the mid-1950s, London re-established itself as the world's premier financial centre and the base of innovative and dynamic financial markets, domestic banking was not so much allowed to slumber as half suffocated. The British experience would appear to provide much evidence for 'free bankers' scepticism about state intervention in the financial sector.

Official policy played its part in the loss of market share by the clearers, and the extraordinary penetration of the British banking market by foreign banks. Any assessment of the impact elsewhere in the economy remains problematic.

The lack of price competition and dampened product innovation may have been unfortunate, but alternative funding sources for industry were available. The banks might have improved the performance of British business by becoming involved in managerial decisions, but the possibility remains that they might have made it worse. It is unclear whether the policy regime applied to international banking would have been so beneficial to domestic banking, as the markets had different characteristics. When financial deregulation came to British domestic banking in the 1980s, the results appear to have been ambiguous. Although the pace of innovation and competitive vigour grew, the more competitive credit market led households to hold higher levels of debt in relation to their income. The impact on savings ratios resulted in credit allocation by price rationing rather than official controls. The superiority of this alternative world, which also appears more prone to instability, in enhancing British competitiveness remains to be demonstrated.

Appendix *Market share of deposit liabilities, 1955, 1965 and 1975*

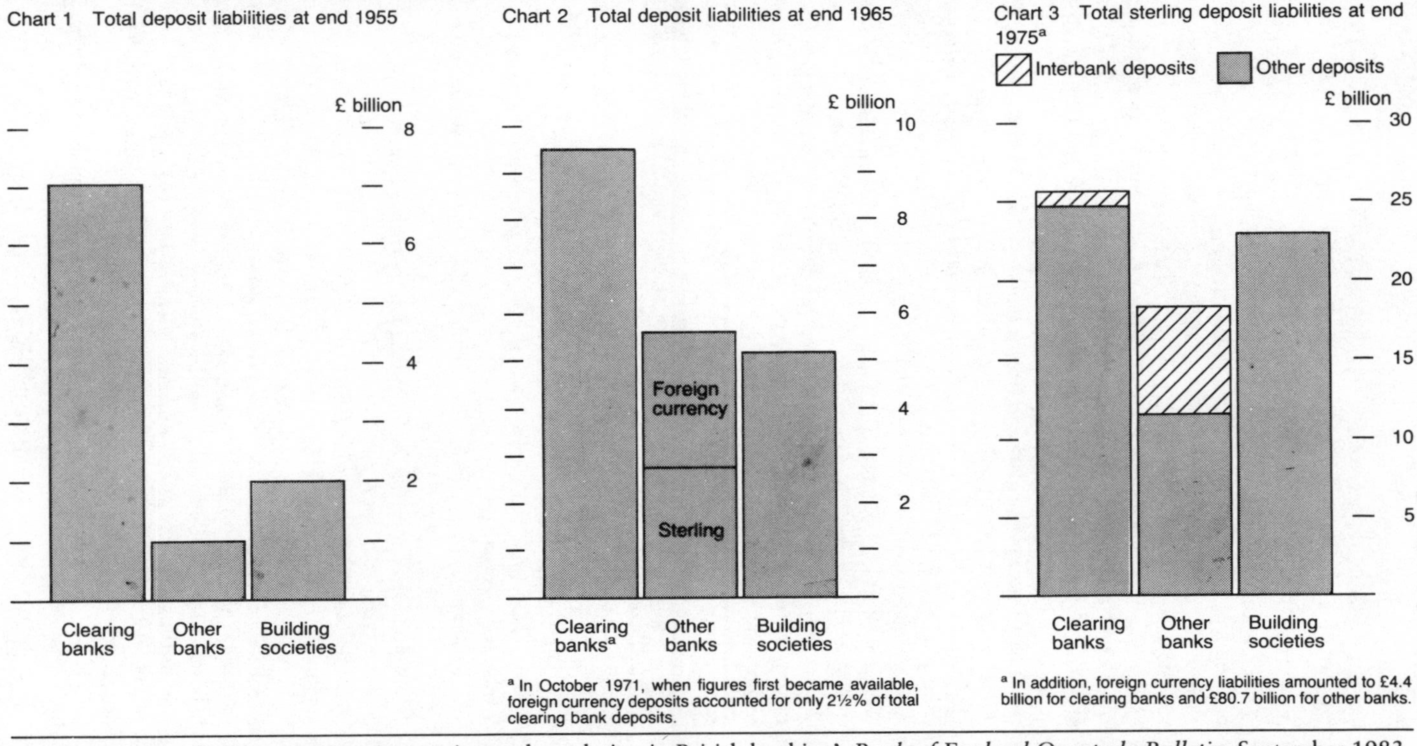

[a] In October 1971, when figures first became available, foreign currency deposits accounted for only 2½% of total clearing bank deposits.

[a] In addition, foreign currency liabilities amounted to £4.4 billion for clearing banks and £80.7 billion for other banks.

Source: J. S. Fforde, 'Competition, innovation and regulation in British banking', *Bank of England Quarterly Bulletin*, September 1983.

Notes

I would like to thank Mark Casson, Mike Collins, Leslie Pressnell, Brian Quinn and Duncan Ross for comments on an earlier draft of this chapter.

1 See (from very different perspectives) S. Pollard, *The Wasting of the British Economy*, London, 1982; G. Ingham, *Capitalism Divided*, London, 1984; E. H. H. Green, 'Rentiers versus producers?', *English Historical Review*, 1983; *idem, The Influence of the City over British Economic Policy, 1880–1960*, Conference on Finance and Financiers in Europe, Geneva, 5–7 October 1989.

2 J. R. Winton, *Lloyds Bank 1918–1969*, Oxford, 1982; A. R. Holmes and Edwin Green, *Midland — 150 Years of Banking Business*, London, 1986; Charles W. Munn, *Clydesdale Bank*, Glasgow, 1988; Forrest Capie and Ghila Rodrik-Bali, 'Concentration in British banking, 1870–1920', *Business History*, 1982; Forrest Capie, 'Structure and performance in British banking, 1870–1939', in P. L. Cottrell and D. E. Moggridge (eds), *Money and Power*, London, 1988; *idem*, 'The evolving regulatory framework in British banking', in Martin Chick (ed.), *Governments, Industries and Markets*, Aldershot, 1990; M. Collins, *Money and Banking in the UK: A History*, London, 1988.

3 For Hayek's views, see for example F. A. Hayek, *Denationalisation of Money — the Argument Refined*, Institute of Economic Affairs, London, 1978. For an extended critique of central banks, see Vera C. Smith, *The Rationale of Central Banking*, London, 1936, and Lawrence H. White, *Free Banking in Britain*, Cambridge, 1984.

4 Collins, *Money and Banking*, pp. 207–11, 398–404; 'London clearing banks: mergers and acquisitions, 1918–67', Appendix 3 to Monopolies Commission Report on Proposed Merger between Barclays, Lloyds and Martins (1968).

5 Capie, 'Structure and performance', p. 96.

6 Collins, *Money and Banking*, p. 211; B. Griffiths, 'The development of restrictive practices in the U.K. monetary system', *Manchester School*, 1973, pp. 4–8; L. S. Pressnell, 'Cartels and competition in British banking: a background study', *Banca Nazionale del Lavero Quarterly Review*, 1970, pp. 386–91.

7 Munn, *Clydesdale*, pp. 36, 215–16.

8 W. J. Baumol, *Contestable Markets and the Theory of Industry Structure*, 1982.

9 Winton, *Lloyds*, pp. 46–7. On inter-war branch expansion in general see Collins, *Money and Banking*, pp. 205–7.

10 Holmes and Green, *Midland*, pp. 167–71.

11 Munn, *Clydesdale*, p. 216.

12 Evidence of Sir W. H. N. Goschen, 7 February 1930 and M. J. W. Beaumont Pease, 13 February 1930, to Committee of Finance and Industry.

13 Steven Tolliday, *Business, Banking and Politics*, Harvard, Mass., 1987, p. 179. J. Foreman-Peck, 'Exit, voice and loyalty as responses to decline: the Rover Company in the inter-war years', *Business History*, July 1981, p. 204.

14 J. H. Bamberg, 'The rationalisation of the British cotton industry in the interwar years', *Textile History*, Spring 1988, p. 86.

15 R. S. Sayers, *The Bank of England 1891–1944*, Cambridge, 1976, I. pp. 249–58.

16 Holmes and Green, *Midland*, pp. 181–5.

17 Collins, *Money and Banking*, pp. 247–55; Edwin Nevin and E. W. Davis, *The London Clearing Banks*, London, 1970, pp. 162–7.

18 Holmes and Green, *Midland*, p. 219.

19 Winton, *Lloyds*, p. 147.

20 Collins, *Money and Banking*, pp. 440–2.

21 Holmes and Green, *Midland*, chapter 10.

22 Quoted in Monopolies Commission Report, 1968, p. 6.

23 J. S. Fforde, 'Competition, innovation and regulation in British banking', *Bank of England Quarterly Bulletin*, September, 1983, p. 364.

24 Collins, *Money and Banking*, p. 202.

25 *Ibid*, p. 411.

26 Report of the Committee to Review the Functions of Financial Institutions 1980, Cmnd. 7937, pp. 67–8.

27 Michael H. Best and Jane Humphries, 'The city and industrial decline', in B. Elbaum and W. Lazonick (eds), *The Decline of the British Economy*, Oxford, 1986, p. 236; Charles Harvey, 'Old traditions, new departures: the later history of the Bristol & West Building Society', in C. Harvey and J. Press (eds), *Studies in the Business History of Bristol*, Bristol, 1988, pp. 246–8.

28 Janet Kelly, *Bankers and Borders: The Case of American Banks in Britain*, Cambridge, Mass., 1977; Collins, *Money and Banking*, p. 411; Committee to Review the Functions of Financial Institutions, 1980, pp. 68–9 and Appendix 3, pp. 416–17.

29 Capie, 'Evolving regulatory framework'.

30 Green, 'Influence of the City'.

31 Evidence of the Committee of London Clearing Bankers to the Committee to Review the Functions of Financial Institutions, November 1977.

32 Peter Hall, *Governing the Economy*, Cambridge, 1986, pp. 61, 92.

33 Sayers, *Bank of England*, II, pp. 552–60.

34 This point is discussed in Griffiths, 'Development of restrictive practices', pp. 11–15 and Capie, 'Evolving regulatory framework'. For developments in the 1960s see Evidence of the Committee of London Clearing Bankers, 1977, p. 212; Winton, *Lloyds*, p. 178.

35 Sayers, *Bank of England*, I, pp. 236–43.

36 Griffiths, 'Development of restrictive practices', pp. 8–11; Pressnell, 'Cartels and competition', pp. 384–5; Collins, *Money and Banking*, pp. 219–20.

37 Sayers, *Bank of England*, I. pp. 236–43.

38 Winton, *Lloyds*, pp. 194–200.

39 Geoffrey Jones, 'The British government and foreign multinationals before 1970', in Chick, *Governments, Industries and Markets*.

40 Holmes and Green, *Midland*, pp. 227–8; Munn, *Clydesdale*, p. 287; Winton, *Lloyds*, pp. 162–3.

41 Unpublished research by Duncan Ross.

42 Munn, *Clydesdale*, pp. 228, 280–1.

43 Collings, *Money and Banking*, p. 419; Evidence of the Committee of London Clearing Bankers, 1977, pp. 68–70.

44 Evidence of the Committee of London Clearing Bankers, 1977, pp. 69–70.

45 Richard Roberts, 'How the City put its house in order', *Financial Times*, 4 November 1989.

46 Capie, 'Evolving regulatory framework'; H. Curtis Reed, *The Preeminence of International Financial Centres*, New York, 1981.

47 Collins, *Money and Banking*, pp. 374–6; Kelly, *Bankers and Borders*, chapter 4; Committee to Review the Functioning of Financial Institutions, pp. 68–9 and Appendix 3, pp. 507–9.

48 Pressnell, 'Cartels and competition', p. 386.

49 Winton, *Lloyds*, pp. 147–8.

50 Capie, 'Structure and performance', pp. 96–8.

51 Pressnell, 'Cartels and competition', p. 383.

52 Geoffrey Jones, 'Lombard Street in the Rivieria: the British clearing banks and Europe 1900–1960', *Business History*, 1982.

53 *Idem*, 'Competitive advantages in British multinational banking since 1890', in Geoffrey Jones (ed.), *Banks as Multinationals*, London, 1990; Winton, *Lloyds*, p. 30.

54 Holmes and Green, *Midland*, pp. 189–92, 213–8.

55 Munn, *Clydesdale*, pp. 305–6.

56 *Ibid*, p. 223; Evidence of the Committee of London Clearing Bankers, 1977, pp. 31–2. For the 'general belief' that 'Barclays was the most efficient, or one of the most efficient, clearing banks', see Monopolies Commission Report, 1968, p. 33.

57 Griffiths, 'Development of restrictive practices', p. 15.

58 Capie, 'Evolving regulatory framework'.

59 N. Crafts, 'The assessment: British economic growth over the long run', *Oxford Review of Economic Policy*, Spring 1988, p. v. This is the central argument of Best and Humphries, 'The City and industrial decline'.

60 Evidence by the Committee of London Clearing Bankers, 1977.

61 Fforde, 'Competition, innovation and regulation', pp. 370–1.

62 Collins, *Money and Banking*, pp. 416–21; Evidence by the Committee of London Clearing Bankers, 1977, pp. 372–4.

63 Collins, *Money and Banking*, p. 419.

8 *Oliver M. Westall*

The assumptions of regulation in British general insurance

1 Introduction

The increasingly competitive and integrated environment of international trade is drawing attention to national differences in industrial regulation. These can raise costs or reduce domestic competition and thereby inhibit international performance. The internationalisation of financial markets has made such issues particularly relevant. Governments have always taken a close interest in them, because of their significance for economic policy and stability and the need to guarantee the security of consumers' assets. These different motives have caused complications, however, for governments have often had difficulty in reconciling conflicting objectives. This is perhaps best recognised in banking, dealt with elsewhere in this collection, where the aims of prudential management, competitiveness, and policy control have not always been easily integrated.[1]

British financial services are often thought to have benefited from a liberal regulatory regime which has contributed to the success of the City of London as an international financial centre. Recent attempts to tighten control have been opposed, partly because it was argued that it would erode the City's comparative advantage. When legislation has been introduced, it has relied heavily on self-regulation, and attempts to tighten external control have been strenuously resisted.

Insurance has shared these assumptions. Michael E. Porter has recently attributed its international competitive success to a *laissez-faire* regulatory environment, possibly the least interventionist in the world.[2] In so far as regulation forms a barrier to trade in services, the insurance business has continued the tradition of 'free trade imperialism', advocating minimum regulation at home and abroad, in order to expand its own potential. All has not been straightforward in recent years, however. Around 1970 relations between government and insurance went through a profound crisis. In 1971 the collapse of a motor insurance company, the *Vehicle and General*, led to serious losses for policy-holders and a collapse of public confidence in government insurance regulation, requiring the introduction of a substantial body of new legislative control.[3] Then in 1972 the Monopolies Commission recommended that the collusive arrangements in fire insurance should be ended.[4] This proposal was not fully implemented until 1985, but it effectively brought to an end practices fundamental to the business for more than a century.

The proposition of this essay is that these two episodes both arose from the same underlying turning-point in relations between government and insurance. They exposed long-standing inconsistencies in the assumptions on which they had been based, which could no longer be camouflaged because of structural changes in the business and in the attitudes of government. The following account will be restricted to a discussion of general insurance for the different competitive process in life assurance raises issues that would require a different analysis. Regulation in general insurance developed from that devised for the life market, however, so it will be necessary to trace the legislative framework from the original Life Assurance Companies Act of 1870.

2 The origins and assumptions of British insurance regulation

Regulation is usually justified in terms of economic welfare to correct market breakdown arising from market power, externalities, or imperfect information. All these have had pertinence for insurance and have been variously addressed by legislation in many countries. Most, however, including Britain, have placed the greatest emphasis on the more fundamental requirement of contract fulfilment. Insurance premiums are paid before claims are met, so if a company becomes insolvent during the term of a policy, the holder has no recourse and may suffer loss if a claim arises.

The reason for this regulatory emphasis lies in the political economy of regulation. The social losses from other sources of market breakdown may be larger than those arising from insolvency, but they are usually more difficult to focus sufficiently effectively to lead to legislation. Market power may lead to excessive premium rates; poor information may mean policies offer inadequate cover; but the resulting costs are difficult to appraise because they are intangible, technical, and difficult to resolve without insurance expertise. Losses may also be diffused among so many policyholders that it pays none to support concerted action. Insurers may oppose interference which threatens their profitability.[5]

By contrast, the issue of solvency is likely to be what recent students of regulation have termed 'salient' and lacking in 'complexity'. The losses bear heavily on an identifiable group who suffer an easily understood and clearly established injustice. Such circumstances are likely to influence opinion and legislator. Established insurers are also likely to support action to protect the reputation of their market, especially if this is shaped to suit their own interests.

These suggestions are consistent with the origins of insurance legislation in Britain. Despite pressure from insurers to introduce control from the early 1850s, when competition in life assurance intensified, it was only after the collapse of the *Albert Life Assurance Company* in 1869 that legislation was introduced. The company caused loss for policy-holders from twenty-four companies that it had taken over, and one contemporary commentator suggested that some £10 million had disappeared.[6] The most important provisions of the resulting 1870 Life Assurance Companies Act (alongside various technical requirements concerning policy-holders' rights and accounting provisions) were that any company starting life assurance should deposit £20,000 with the authorities, returnable when life funds reached £40,000. All companies should also provide validated accounts, supported by a periodic actuarial appraisal of life business, which were returned to the Board of Trade for publication.[7] Thus was laid down the cautious mid-Victorian 'nightwatchman state' regulation, which relied upon the principle of 'freedom with publicity' that was to be the main element in regulation for a hundred years. No positive role was given to government. The deposit was designed to prevent frivolous or fraudulent entry. Otherwise the onus lay entirely with consumers to ascertain the financial status of companies from published accounts.

The legislation was almost unchanged when it was extended to most forms of general insurance in 1907 and 1909.[8] The earlier act was justified by the external consequences of a private contract. Employees whose employers were insuring against the liabilities imposed by Workmans' Compensation legislation, needed to be protected against the failure of a policy contract to which they were third parties. The wider extension to fire and accident insurance in 1909 was represented as necessary because of the recent failure of fraudulent companies abroad which had tarnished the international reputation of British insurance. Further research may discover this to be an example of regulatory 'capture', however. Established companies were facing severe competition from new concerns, attracted by rapid growth in accident insurance and its marketing coalescence with fire insurance. The new non-returnable deposit of £20,000 must have been extremely convenient for them.[9]

While the main principles of regulation remained largely unchanged, problems in the inter-war years led to some legislative evolution. Marine insurance failures caused by that market's contraction after a wartime boom, affected other markets. An official inquiry reporting in 1927 recommended that it should be regulated, though no immediate action was taken.[10] The most serious problems arose from the growth of motor insurance. In 1930 it was identified as a separate class to be reported and requiring a separate deposit. When its competitive volatility led to several serious failures in the 1930s, legislation in 1933 and 1935 granted the first positive regulatory powers by enabling the Board of Trade to wind up insolvent companies. Compulsory motor insurance was imposed in 1930 to provide compensation for motor accident victims. Some motorists simply sought the cheapest policy, raising again the issue of the consequences for the third party if the company collapsed. An official committee was established in 1936 to find a solution to the issues raised by the externalities of compulsory insurance. Its report in 1937 was incorporated into post-war agreements between government and insurers.[11]

In 1946 a new Insurance Companies Act made some attempt to deal with these problems.[12] Marine insurance was brought within regulatory control. A more sensible approach was made to the issue of solvency. The deposit requirements of the 1909 Act were abolished. The sum required had borne no relationship to the liabilities incurred by large companies. It was replaced by a requirement that companies have a paid-up capital of £50,000 and that the margin between their assets and liabilities should be the larger of that amount or 10 per cent of their previous year's premium income. Protection for unsuccessful claimants under the compulsory motor third party cover was provided by a voluntary agreement between government and motor insurers. The latter funded the *Motor Insurers' Bureau* which would meet any claim arising under compulsory cover that was not adequately paid by a company. No further legislative changes were made until 1967.[13]

Insurance regulation in Britain therefore remained essentially passive, relying on publicity to alert consumers and constrain companies. While the Board of Trade was eventually given powers to wind up companies, these were infrequently used, and then usually for insignificant concerns. The government acted as undertaker rather than ringmaster, and this was in contrast to regulatory regimes in most other countries with insurance industries of similar size and sophistication.

In the United States, the most obvious and important example, solvency was monitored as early as 1799.[14] Many states followed New Hampshire, which established the first insurance regulatory board in 1851, in licensing agents, setting reserve requirements and monitoring company operation. Individual state commissioners formed a national organisation in 1871 to co-ordinate the drafting of regulatory legislation. Many imposed standard policy terms from 1873. During the progressive era corrupt insurance management provided rich pickings for investigators, leading to new regulations governing investment policy. The failure of so many companies after the San Francisco fire in 1906 focused attention on fire insurance. Official inquiries concluded that rate regulation was necessary to ensure solvency. Legislation allowed rate bureaux to propose premium rates which had to be filed with the authorities to ensure that they were 'reasonable and adequate'. Some idea of the active nature of control can be deduced from the 179 staff that the New York Insurance department employed in 1915.[15]

By contrast the staff of the British Board of Trade responsible for supervising all British insurance companies included eight members in the early 1960s.[16] Collusion between state governments and insurers allowed the creation of profitable cartels in the inter-war years, until important legal decisions in the early 1940s re-established the applicability of federal anti-trust legislation to insurance. By 1951 new standardised legislation was introduced in most states specifying that all premium rates should be filed and approved by insurance commissioners as not being 'inadequate, excessive, or unfairly discriminatory'. Rating organisations had to be state licensed and variations from bureau rates were only allowed if they were filed and defended by the originating company. Competitive forces undermined this system in the 1960s and few rate reductions were refused by insurance commissioners. In the early 1970s, however, inflation reinvigorated rate control as consumer groups used the legislation to limit premium rate increases.

European countries have adopted similarly interventionist regimes. In Germany insurers required government licenses from 1901 and these were used to control entry, allowing cartels to develop which controlled premium rates. When these collapsed in the inter-war years, the government intervened to impose uniform premium rates, control agency commissions and standard contracts. Switzerland followed a similar pattern from its initial legislation in 1885. Its regulatory agency approves premium rates after negotiation between consumer groups and insurers. In Sweden a supervisory board for insurance was established in 1904. Later legislation established government control to ensure 'reasonable' margins on premium rates and no cross-subsidisation between classes of insurance. Entry was also regulated to ensure that new companies did not affect the 'sound' development of the insurance market. In France regulation appears interventionary in form, providing for the licensing of insurers, the monitoring of solvency, the control of the disposition of investments, and the power to set maximum and minimum premium rates, though apparently the latter is not used actively.[17]

The essential characteristic of these other regimes is that they have moved beyond the passive approach of information provision to a more active role of attempting to create an environment that will make insolvency unlikely. Publication of the financial status of companies may provide advance warning of difficulty and constrain companies from continuing to operate when insolvent, but its usefulness is limited. Consumers may not obtain the requisite information or be able to interpret it correctly. Reporting is always delayed and information relevant to insurance insolvency is sometimes technical and inadequately revealed in accounts. But of greatest importance, the passive approach makes no more than an indirect impact on the particular root causes of insolvency in general insurance.

These arise in part from the potentially fragile nature of price formation.

Competition is always likely to focus on premium rates because of the limited opportunities for differentiating insurance indemnity. Markets are usually characterised by an oligopolistic interdependence that is strong enough to ensure great sensitivity to rate competition, but too weak to ensure successful implicit collusion. Thus when competitive pressures arise, they will be expressed through rate reductions which lead quickly to price warfare, with rates moving down dramatically as companies try to protect their business. This process may be sustained by the receipt of premiums before costs which maintains liquidity. In practice this may lead to the erosion of long-term solvency because of the temptation to cut rates below costs, to maintain the life preserving flow of revenue. Weak companies or managements are likely to fall into this position in the hope that something will turn up. Publication of accounts has little effect because it is delayed and the estimation of the real liabilities attaching to an underwriting account is probably only known to those managing it. For these reasons, price warfare can precipitate insolvency very quickly. This is why many countries have seen insolvency as best avoided by the creation of market stability through rate regulation or entry control.

Regulation of this kind, however, inevitably interferes with the competitive process that tends to ensure efficiency, encourage innovation, and limit profitability. This is why insurance interests have often been among the most enthusiastic advocates of regulation. Though it is too simple a view of insurance regulation as a whole, Stigler's 'capture' approach has certainly been an important element in the spread of interventionist legislation.[18] This conflict with the public interest in limiting market power has not been unrecognised, as the account of American legislation above suggests. Rate regulation often incorporates the need for 'reasonable' rates that tread the narrow path between being so low that they threaten insolvency, and so high that they allow monopoly profits. Rate regulation can certainly work in both directions, as its use by consumer groups in the USA in the 1970s shows. In other cases rents earned through market power have been taxed in other ways, as by the corrupt political machine in Kansas in the 1930s.[19]

Implicit in the British approach to regulation has been the assumption that its extension into more direct forms of intervention would reduce the effectiveness of competition in avoiding these problems of market power. Interference with business was seen as inevitably reducing consumer welfare. This was the characteristic form of British self-delusion. Competition was advocated in principle, but constrained in practice, in the interest of a powerful and persuasive business interest. Insurance regulation remained simple in form and passive in action because it assumed the existence of a system of market self-regulation that made government intervention along the lines practised elsewhere unnecessary. At the same time, it sanctioned a level of market power whose cost to consumers was never appraised. It was a classic pragmatic compromise which was approved by all because it seemed to work — as long as no one questioned its operation and assumptions too closely.

Market regulation in general insurance was formalised when the Fire Offices' Committee (FOC), whose origins stretched back to the 1790s, was established in 1868.[20] Through the late nineteenth century it erected an increasively pervasive and complex series of 'tariffs' which controlled premium rates, policy conditions and commission terms in all the main classes of fire insurance. A parallel organisation, the Accident Offices Association (AOA) was formed in 1906 to perform a similar role in accident insurance, including motor insurance which expanded so rapidly after 1920. Technical factors, including control over essential reinsurance arrangements and comprehensive underwriting data, gave the FOC a strong hold over the fire insurance market. For long periods it was practically impossible to operate on a large scale in it without tariff membership. The AOA operated similar arrangements, and although these were not so effective as in the fire market, the required support of all FOC members buttressed its tariffs.[21] Behind the protection of these restrictions on competitive activity, the characteristic modes of British insurance activity emerged. Competition was redirected into non-rate marketing, most obviously through the creation and expansion of branch office organisations. The goodwill and overhead costs implicit in these created barriers to entry which strengthened the position of both tariffs.[22]

When it was successful, the British insurance tariff system created stable markets, protecting consumers and insurers from the potential for instability and insolvency described above. Premium rate control prevented below cost rate cutting which could threaten financial stability. The pooling of underwriting data in the standard tariff classes of risk enabled members to quote premium rates that were reliably based on the best evidence available. Its supporters argued that this also enabled it to quote rates that were equitable between insurer and insured and between risks. In these ways the British tariff organisations fulfilled the objectives sought by more interventionist legislation elsewhere. When the tariff was working well it was difficult for a company to fail.

Of course, the tariff can be seen in a different light. The more effective it became, the more likely it was to exploit its market power to raise rates. It had little incentive to rate according to experience if it could get away with higher margins, especially when aggregate demand remained relatively inelastic. Furthermore, its existence redirected competitiveness into the expansion of branch office organisations and other forms of marketing, and away from directions that might have been more genuinely worthwhile for consumers. Its control was exercised over policy terms as well as prices, thus discouraging innovations that might broaden the market and benefit consumers.

The main threat to its activities came from entry. In fact, there were nearly always insurers operating outside the tariff, in theory creating some competitive pressure to restrain margins. Certainly most new forms of policy cover in both the fire and accident insurance markets were initiated by independent underwriters, particularly those at Lloyd's. In practice, however, most of them followed the premium rates set by the tariff and had as much interest in its survival as the tariff members themselves. They would never have been able to compete with the large tariff companies if the latter had competed openly by price, product, and marketing organisation. Their existence therefore only partially moderated the competitive control of the tariff companies. It was only when small, weak, or poorly managed firms entered the market that their aggression could threaten stability. This is why there must be a suspicion that its members were always inclined to support legislative control over entry, as in the introduction of the 1909 Act when so many new accident companies entered the market.[23]

As a system designed to avoid insolvency, this self-regulatory system was successful in fire insurance for there were no significant failures in that market. In accident insurance the tariff system came under greater pressure because entry and growth were easier outside the tariff system. This could lead to difficulty, as in the 1930s, when serious rate cutting occurred.[24] Significantly, this led to the important extension of legislation to allow the Board of Trade to intervene to close companies down.[25] But for the most part the accident tariff held together. Of course, with the implicit collusion of independent insurers, the success of the system in securing the alternative goal of public policy of limiting market power must remain more doubtful. This, however, does not appear to have been a 'salient' issue. It was widely assumed that if government did not intervene, competition would resolve the problem.

However, the assumptions and structures on which this apparatus of control was based were challenged in the late 1960s. In October 1968, the year in which the FOC celebrated its centenary, two events encapsulated the change. The Monopolies and Mergers Commission was asked to investigate the supply of fire insurance, thus calling into question the balance of public interest between rate regulation and market power.[26] In the same month, ten leading members of the AOA proposed that the accident tariffs be ended and they were terminated from the start of 1969.[27] This led to severe premium rate competition which demonstrated the extent to which the old regulatory framework had depended on the existence of the tariffs by precipitating the most dramatic insolvency in twentieth-century British insurance.

3 The end of the accident tariff

The motor insurance market had never been comfortable for the tariff companies. The rapid growth in vehicle ownership, the large premiums necessary to cover claims, and the almost universal purchase of policies, generated a large and buoyant market, both between the wars and after 1950. It could not be ignored by established insurers, for the revenue financed larger branch organisations and reserve funds, both of which were of competitive importance in all insurance markets. Yet high premiums and the regular exchange of vehicles made the market rate competitive, allowing independent insurers to win business by cutting rates. The AOA found it difficult to restrict such entry because reinsurance and pooled underwriting data were less important in motor insurance. They had to rely more on marketing strength and the capacity to subsidise low profitability from more profitable underwriting or investment earnings in other markets. This had some success because rising claims and the expenses of transacting the business continually squeezed profitability.[28]

Despite this the tariff steadily lost business to independents, but this did not normally disrupt the stability of the market. Most independent underwriters had as strong an interest in the survival of the tariff as AOA members. They simply shaded their rates sufficiently below the tariff to grow, but without forcing the AOA to reduce rates, and thus provided profitability for all. Stability was only threatened by independents who sought more rapid growth. This was what had occurred in the 1930s when new firms with fragile finances had cut rates severely to try to establish a market position. Temporary AOA rate reductions and rising claims had forced them out.[29]

Difficulties intensified in the post-war years. Motor insurance encountered greater sensitivity to the level of premium rates as lower income groups bought cars. Rates were forced up by the inflationary pressures on labour intensive repair costs. The search for competitive rates was facilitated by an explosion in the number of small High Street brokers, of whom there were some 5–6,000 by the late 1960s.[30]

Established independent insurers at Lloyd's and companies such as the *General Accident* and the *Eagle Star* maintained a steady pressure on the AOA's market share through the 1950s. Their success, however, encouraged the entry in the early 1960s of new concerns. Some were clearly fraudulent, yet their dramatic success demonstrated the unthinking sensitivity of the market to rate reductions. The *Fire Auto and Marine*, run by the notorious Emile Savundra, collapsed in 1966 after attracting 200,000 policy-holders in its short life, despite (or because of) its clearly inadequate premiums at half the market level. This company and others disappeared quite quickly, but some, whose background was apparently less dubious, such as the *Vehicle and General* and the *Midland, Northern and Scottish* presented a more sustained challenge.[31]

The success of these incursions can be largely attributed to the inherent inflexibility of the AOA. Through the 1950s and 1960s it raised its rates several times to cover increasing costs. But it retained a rating structure based largely on the value and power of cars that had not fundamentally changed since the establishment of the motor tariff before the First World War. This ignored evidence that motorists were more important determinants of claims than vehicles. The AOA's inflexibility allowed independent companies to discount tariff premium rates selectively (mainly through larger no claims discounts) yet remain profitable by selecting motorists with superior claims experience. As a result, the tariff share of the home market fell from two-thirds of the domestic market in the early 1950s to one-third in the mid-1960s.[32]

This ineffective response was inherent in the AOA's organisation. Its procedures over-represented less efficient companies who were incapable of successful competition, but more effective tariff insurers could not ignore them without threatening the AOA's control. As tariff companies lost business, their underwriting increasingly reflected the experience of below average motorists, reducing margins and making reductions seem impossible. Costs imposed by inter-tariff branch office competition were also a problem.

By 1964 it had become clear that motor insurance required radical reappraisal. Significantly, it was the British Insurance Association (BIA), representing all reputable British insurers, including independents, that initiated a new approach. McKinsey's were commissioned to investigate. Their 1965 report, though too controversial to publish, is thought to have proposed that many details previously considered important in motor underwriting could be discarded to simplify procedures. They also suggested shared facilities to control repair costs, and different commissions to reflect the service provided by part-time agents and full-time brokers.[33]

The AOA's reaction to McKinsey was announced as a 'New Deal for Motorists' in 1966.[34] This altered the balance of premium rates in favour of better motorists through a preferential discount for those with sustained claims free records, which was retained after one claim. This concession was balanced, however, by penalising less attractive motorists with large rate increases to maintain revenue. As these motorists now formed the bulk of tariff companies' business, this provided an opportunity seized by the more aggressive independent underwriters. The *Vehicle and General's* motor premium income soared from £6.9 million in 1966 to £12.8 million in 1967; less dramatically, Lloyd's home motor business rose from £35.6 million to £40.5 million.[35] A revolutionary attempt to put the full recommendations of the McKinsey Report into practice was mounted through the 'Alpha' scheme of the *Midland, Northern and Scottish* which based premiums almost entirely on the age and experience of motorists, reducing administrative costs.[36] As the consequences of the 'New Deal' for market share and profitability became apparent through 1968, the larger tariff concerns took stock and in October ten proposed that the AOA tariff be terminated and it ceased to operate from the end of that year.[37]

With the end of the tariff, its former members quickly sought new ways to compete successfully. There was a proliferation of new policies that were highly differentiated through branding, advertising, and policy conditions. Rates for preferred motorists were also cut below the 'New Deal' scheme. While they wanted such risks, this was strategically motivated. The larger ex-AOA members' chief advantage over the new aggressors lay in their profitably diversified underwriting in other insurance markets and investment income derived from large reserves. Relying on these, they decided to squeeze the profitability out of motor underwriting.[38] In this they were assisted by a rise in claims in 1969, caused partly by accelerating inflation. In 1969 BIA members had underwriting losses of 7.3 per cent on a UK premium income of £210 million; in 1970 this rose to 14.5 per cent on £215.7 million.[39] As competitive pressure intensified, insurers played 'chicken', waiting for the first to raise rates and lose business. In 1969 the 'Alpha' policy was forced out of business.[40] In 1969 the *Vehicle and General's* business had grown to some 8 per cent of the UK motor insurance market, but in 1970 it was forced to raise rates twice, once by 35 per cent. Eventually, deprived of the revenue growth that had financed expansion, it was forced into liquidation in March 1971, leaving 800,000 policy-holders without cover or recourse for claims.[41] Immediately, many motor insurers raised their rates by between 15 and 37.5 per cent.[42] Market control had been imposed by financial muscle at the cost of the most dramatic modern British insurance insolvency.

The collapse of the *Vehicle and General* was on an unprecedented scale. By disrupting market equilibrium the growth of the *Vehicle and General* precipitated the end of the tariff and its own demise. While it was undoubtedly mismanaged to the point of fraud, its ability to continue operating on a scale that involved so many in its failure exposed the weakness of British insurance regulation. Criticism was directed at the responsible Department of Trade and Industry officials and their relations with the British Insurance Association, but the problem was more fundamental than that. The collapse of the tariff in motor insurance pressed the passive regulatory framework to breaking point. It was the weakness of the tariff system in the 1960s that allowed such extensive entry and then its collapse which led to the failure of the *Vehicle and General*.[43]

The motor insurance market was always the weak point in the tariff system and therefore forced changes in the regulatory regime. Motor insurance failures in the 1930s led to early concern about the effectiveness of regulation. The collapse of *Fire Auto and Marine* in 1966 led to changes in the insurance provisions of the 1967 Companies Act which tightened control by allowing the Board of Trade to refuse the authorisation of insurers, control operation, examine reinsurance, require information from companies, and exclude individuals from company management. The failure of the *Vehicle and General* led to a series of more radical legislative changes, mostly consolidated in the 1974 Insurance Companies Act.[44] The principal innovations included a detailed vetting of new entrants who obtain official approval of a detailed business plan and the submission of quarterly accounts for ten years. The DTI can intervene in the running of any company whose solvency is doubtful and can prevent investment in certain assets or enforce their realisation. Finally the department holds residual powers to 'take such action as appears ... to be appropriate for the purpose of protecting policy-holders ... against the risk that the company may be unable to meet its liabilities ...'[45] These new powers were supported by a different attitude on the part of the DTI which increased the number and expertise of its officials. Direct regulation was supported by the 1975 Policyholders' Protection Act which guarantees policy-holders 90 per cent of the benefits promised under their policies (100 per cent in the case of compulsory insurance) met from levies on other insurers in the market.[46] Brokers have also been controlled through a registration system, undoubtedly partly because of their poor performance in not warning policy-holders about the risks in insuring with a company like the *Vehicle and General*, not unrelated to the large commissions it offered.

In the 1970s, therefore, the regulatory environment became far more interventionary mainly in order to compensate for the consequences of the end of the tariff. While legislation did not allow it to take over the rate regulation that had been the main contribution of the AOA tariff to the relative stability of the motor insurance market, its discretionary powers were so wide-ranging, especially in relation to entry, that it can be assumed that the market was protected from the type of aggressive rate cutting that had led to destabilisation and insolvency in the past. 'Freedom with publicity' was replaced by 'freedom with responsibility' and the DTI was to be the arbiter of responsibility.[47] Perhaps established insurers found that legislation was a better way to achieve the objectives they had previously sought through the AOA.

4 The end of the fire tariff

In the fire insurance market the tariff system was strengthened in the 1960s, but this brought government intervention from another direction which created problems for both insurers and the authorities. During the 1950s the FOC operated in as lethargic a way as the AOA. Loss experience was not collected systematically and the level and structure of fire tariffs had scarcely changed since the 1920s.[48] Yet there had been great changes in fire hazard with the spread of sprinkler protection, automatic fire alarms, telephones, and mechanised fire engines. Tariff companies had therefore benefited from fixed premium rates and falling losses.

The potential this offered was not lost on independent underwriters. From the First World War they had mounted an increasingly powerful attack on the tariff. Lloyd's had played an important role, winning direct business through its brokers, and by reinsuring independent companies.[49]

The tariff had responded with two restrictive devices. In 1930 it had introduced the 65/35 rule, whereby its members would only participate in the co-insurance of large risks, requiring several insurers, if independent underwriters were restricted to 35 per cent.[50] A tariff office facing independent competition could also apply to the FOC for a specially reduced rate.

None the less the tariff companies lost market share through the 1950s. The main reason was the expansion of large brokers, who competed fiercely to win clients. By the 1960s there were some 2,000 such firms, to be distinguished from the small brokers who sold motor insurance.[51] They sought lower premiums by placing risks with independent insurers or by persuading tariff insurers to seek a rate reduction. By the late 1950s the FOC was granting about a thousand special rates each year.[52] Behind its apparently secure facade the tariff system was being eroded by competition.

It was reprieved by a dramatic change in loss experience. In 1958 British domestic fire losses had been £24 million, broadly representative of preceding years. In 1959 they rose to £44 million, and steadily increased to £100 million in 1968.[53] Technological change was the most important reason. As the petrochemical and plastics industries expanded, their hazardous processes assumed greater importance in fire hazard. The spread of assembly line production increased the number of large industrial buildings with no internal divisions to prevent the spread of fire. The increasing importance of electrical and electronic goods, both as products and industrial equipment, created the possibility of extremely large losses from small fires.[54]

Sustained losses forced the home fire insurance market into unprofitability for the first time in underwriters' memories. Through the early 1960s a fierce battle was joined between the brokers and the tariff companies. The pressure became so intense that even the independent offices formed their own 'non-tariff' tariff, to protect themselves from brokers.[55]

This experience shocked the tariff into action. Confidential discussions between the FOC and independent underwriters led to a 15 per cent increase in FOC rates in 1963, which was followed by the independent companies.[56] This was seen, however, as a temporary measure. Retrospective data on loss experience was obtained from tariff members and used to revise tariff rates in each class in the following year. Some remained untouched; others were raised by up to 200 per cent.[57] The independent market, as anxious as the tariff to restore profitability, followed these increases.[58] As profitability had fallen, the availability of reinsurance contracted, restricting the independent companies' competitiveness and providing the FOC with a stronger grip on business. Relations between the tariff companies and the fire insurance brokers were sweetened by a preferential commission for the larger brokers.[59]

It was in these circumstances that the government referred the supply of fire insurance to the Monopolies Commission in 1968. Before 1965 competition policy had been restricted to manufacturing, but the 1965 Monopolies and Mergers Act had deliberately extended it to services.[60] The reasons for the earlier exclusion are not clear, but the 1964 White Paper which preceded the new legislation had suggested that little was known about restrictive practices in services. The same pragmatic approach that had characterised the initial 1948 Act's approach to manufacturing was therefore proposed, allowing Monopolies Commission investigations into service activities, but not applying the full and dogmatic force of restrictive practices legislation.[61] The FOC's market share — 63 per cent of the domestic market in 1968 — was well over the minimum level justifying a reference.[62] Change came before the Commission reported. From the beginning of 1971 the FOC voluntarily abandoned the tariff for private household insurances, removing an indefensible element from its case, to concentrate on the more important issue of commercial and industrial insurances.[63]

The FOC defended its existence on public interest grounds by arguing that stable profits were important to guarantee sufficient capacity for the largest risks; that premium rates should be equitable between risks and encourage policy-holders to reduce fire hazard; that without collusion commission rates would be forced up; and that reasonable rates were guaranteed by competitive pressures from brokers and independent insurers.[64]

The Commission reported in 1972, finding that independent insurers implicitly colluded with the tariff and recommending that the FOC end its rate fixing activities and the 65/35 rule restricting independent insurers' participation on large risks. It suggested, however, that some of its other activities, especially the collection and dissemination of information on claims experience, should be continued and include independent underwriters.[65] The government immediately accepted the recommendation in principle, but the tariff companies resolved to fight the decision as vigorously as possible.[66]

In fact, implementation was delayed until 1985.[67] The reasons for this remain in inaccessible archives but the clues that exist suggest that the recommendation placed the government in a dilemma. Initial delay may be explained by the extensive new insurance legislation in 1973 and 1974, described above, which must have stretched the capacity of official insurance expertise and required the maintenance of goodwill with the business. Then the new Office of Fair Trading, established in 1973, was given responsibility for the implementation of Monopolies Commission recommendations, and this administrative change may have postponed implementation.

However, no action was taken during the five years of the subsequent Labour administration returned in 1974. This must have been a clear political decision, because the OFT could only act on the request of a minister. No public explanation is available, but it seems likely that the issue became intertwined with the concurrent international financial problems. The dramatic collapse of the British security market in 1974 seriously eroded the financial base of many British insurers, threatening their solvency. After the motor insurance catastrophes, the collapse of British fire insurers would have been politically disastrous and harmed the reputation of British insurance abroad, risking invisible earnings when the balance of payments was a major concern. It cannot have been difficult to persuade the government that it was not the moment to precipitate a crisis by ending a market agreement that protected profitability and solvency. The public interest objectives of competition and solvency came into conflict and the long-standing commitment to the latter triumphed. The FOC thus continued to operate much as before. In 1972 it reduced several tariff rates; in 1974 it promulgated the first tariff for the plastics industry. It attempted to introduce a cross-market scheme to collate loss experience on the lines suggested by the Monopolies Commission, but this proved difficult to implement.[68]

However, the competitive structure behind the facade had changed. Mergers in the 1960s had reduced the number of large insurance groups to six or nine, creating a structure more able to rely on implicit collusion.[69] Then, as the stock market recovered from the collapse of 1974, the value of reserves rose, increasing acceptance capacity. Self-insurance became more common among large industrial companies. Some brokers began to insure risks abroad. Competitiveness therefore intensified and with it many practices common in the 1950s were resurrected. The 65/35 rule became irrelevant in many classes of business because larger insurers were able to accept far larger risks without co-insurance. Brokers again forced insurers to bid against one another. The FOC reverted to granting special rates to protect its members' business. Experience rating, whereby premium rates were reduced in the absence of serious claims, became common.

The new Conservative government in 1979 focused quickly on the survival of the FOC. In 1980 the DTI agreed that three or four years would be given to allow an adequate statistical system of reporting cross-market claims experience to be created.[70] Meanwhile several tariffs governing shops and other small commercial premises were ended immediately and other tariffs were gradually phased out. The FOC was finally wound up on 30 June 1985, when its functions other than rate fixing were assumed by the Association of British Insurers, a new successor organisation to the BIA.[71] Perhaps it was appropriate that one of the oldest, most comprehensive and sophisticated of all British restrictive practices should linger on into an age when it assumed almost the status of a museum exhibit. The FOC had become a pale reflection of its former self and the fire market had moved on to new competitive structures that made it largely redundant.

5 Conclusion

This chapter has argued that the light regulatory framework surrounding British general insurance from 1907 (or even before that initial legislation) until the early 1970s, was largely shaped by the existence of relatively effective self-regulation through collusion. For many years this satisfactorily addressed the insolvency problem which was the main issue of public concern, by normally ensuring stable markets. Thus the success of British regulation, which is sometimes attributed to the form of legislation, must be partly attributed to self-regulation. When collusion became ineffective, as in the motor insurance market, or conflicted with the other public interest objective of encouraging effective competition, as in fire insurance, problems arose.

This interpretation raises new issues for investigation. No attention has been paid in this account to any interaction between insurance interests and government in the creation of this regulatory framework. It seemed most important to emphasise the most fundamental issue. However, while it is difficult to discuss Stigler's idea of 'capture' in the context of such modest intervention, the possibility that legislation has been encouraged and shaped by insurers, to support their collusive activities, should be investigated.

It also seems reasonable to suggest that the consumer may have paid a price for the stability that self-regulation created. It is not clear that the tariffs reflected the cost of insurance efficiently, either in aggregate or in the allocation of costs between different types of risks, during the first half of the twentieth century. Their encouragement of marketing organisations imposed other costs whose benefit to consumers was ambiguous. Innovation was also constrained by the control of policy terms as well as premium rates. While the tariff system was challenged by the growth of independent underwriting with increasing success through the twentieth century, the market power of tariff companies proved extremely resilient and effective in resisting equitable premium rates and product innovation. Its existence as a price leader was as important in supporting the profitability of independent underwriters as much as that of its own members.

The effectiveness of self-regulation also has international implications. The success of British insurance abroad since the early nineteenth century, while partly dependent on overseas investment and the formal and informal empire, has often been attributed to a superior competitiveness. The assumption has been that this was based on superior efficiency generated in a highly competitive domestic market, unconstrained by government regulation. If, contrary to the whole thrust of Porter's recent work, this is questioned, alternative explanations must be investigated.[72] Did the capacity to develop overseas business owe something to a profitable domestic market that could finance overseas ventures? Did the difficulty of domestic expansion in a closely controlled oligopolistic market force companies to look abroad for expansion? Competitiveness operates in the other direction as well. There has been a virtual absence of foreign insurance intervention in Britain, outside the specialist City insurance markets. Can this be explained by the barriers to entry created by the powerful marketing organisations built up behind the tariff, which contrast so strikingly with the agency controlled business in most foreign markets? The link between domestic competitiveness and international competitiveness may prove more complicated than sometimes assumed.

Finally, the discussion also raises general issues concerning the idea of self-regulation, which has been so important in the British financial service sector. The case of general insurance seems to show that while self-regulation may resolve some public interest issues very efficiently, it may leave others more open. In particular, it usually depends on barriers to entry or operation of some kind, whether these are inherent in the market, or granted by government intervention. Such barriers obviously raise questions about the competitive efficiency of the market and the extent to which they are being operated in the interest of consumers or producers. Self-regulation cannot therefore be assumed to encourage the maintenance of competitive markets. This conflicts with the usual picture of British financial services as being based on highly competitive markets in the City of London. Perhaps efficient self-regulation and competitive markets are incompatible in this sector — though the trend in recent British legislation has tended to resist this unpalatable possibility.

Notes

1 See Jones's contribution to this volume.

2 Michael E. Porter, *The Competitive Advantage of Nations*, London, 1990, p. 664; see also Julian Tapp, 'Regulation of the UK insurance industry', in Jorg Finsinger and Mark V. Pauly, *The Economics of Insurance Regulation: a Cross-National Study*, London, 1986; the editors' introduction provides international comparisons.

3 *Report of the Tribunal appointed to inquire into certain issues in relation to the circumstances leading up to the cessation of trading by the Vehicle and General Insurance Company Limited* (HL80 HC133 1972); see also Tapp, 'Regulation', pp. 33–5.

4 The Monopolies Commission, *Fire Insurance Report on the Supply of Fire Insurance* (HC396 1972).

5 This discussion is prompted by Kenneth J. Meier, *The Political Economy of Regulation: the Case of Insurance*, New York, 1988, chapter 2.

6 Barry Supple, *Royal Exchange Assurance: A History of British Insurance 1720–1970*, Cambridge, 1970, pp. 137–45; Cornelius Walford, *The Insurance Cyclopaedia*, Volume I, London, 1871, pp. 46–9; H. E. Raynes, *A History of British Insurance*, London, 1948, pp. 354–5.

7 Raynes, *British Insurance*, pp. 355–8.

8 Raynes, *British Insurance*, pp. 358–4.

9 Oliver M. Westall, 'David and Goliath: the Fire Offices' Committee and

non-tariff competition, 1898–1907', in Oliver M. Westall (ed.), *The Historian and the Business of Insurance*, Manchester, 1984, pp. 144–8.

10 H. E. Raynes, *British Insurance*, pp. 364–5.

11 Oliver M. Westall, 'The invisible hand strikes back: motor insurance and the erosion of organised competition in general insurance 1920–1938', *Business History*, xxx, 1988, pp. 432–50; Raynes, *British Insurance*, pp. 364–8; W. A. Dinsdale, *History of Accident Insurance in Great Britain*, London, 1954, chapters xix and xx on legislation; *Report of the Committee on Compulsory Insurance*, 1937, Cmd. 5528.

12 Raynes, *British Insurance*, pp. 368–71.

13 Dinsdale, *Accident Insurance*, chapters xx.

14 Meier, *Political Economy*, passim; Finsinger and Pauly, *Economics of Insurance Regulation*, pp. 65–110.

15 Edwin Wilhite Patterson, *The Insurance Commissioner in the United States: A Study in Administrative Law and Practice*, Cambridge, Mass., 1927, p. 48.

16 *V&G Tribunal*, p. 22.

17 Finsinger and Pauly, *Economics of Insurance Regulation*, passim.

18 G. J. Stigler, 'The theory of economic regulation', *Bell Journal of Economics and Management Science*, Spring 1971.

19 Meier, *Political Economy*, p. 64.

20 On the evolution of the FOC see P. G. M. Dickson, *The Sun Insurance Office, 1710–1960*, London, 1960, pp. 149–60; Supple, *Royal Exchange Assurance*, pp. 127–30, 217 and 282, Hugh A. L. Cockerell, 'Combination in British fire insurance', in F. Reichaert-Facilides, F. Rittner and J. Sasse (eds) *Festschrift fur Reiner Schmidt*, Karlsruhe, 1976, and Westall, 'David and Goliath'.

21 Westall, 'Invisible hand', pp. 434 and 441.

22 Oliver M. Westall, 'The evolution of marketing strategy in the British general insurance market, 1700–1939', in Terence Nevett, Kathleen R. Whitney and Stanley C. Hollander (eds), *Marketing History: the Emerging Discipline*, East Lansing, MI, 1989.

23 Supple, *Royal Exchange Assurance*, pp. 231–7.

24 Westall, 'Invisible hand', pp. 444–5.

25 See footnote 11.

26 *Post Magazine and Insurance Monitor (PM)* 17 October 1968, p. 1950.

27 *PM*, 24 October 1968, p. 1969.

28 This discussion is based generally on: Westall, 'Invisible hand'; H. A. L. Cockerell and G. M. Dickinson, *Motor Insurance and the Consumer*, London, 1980; Ronald Beale, *After the V&G Crash*, London, 1972; and Peter J. Franklin, 'Some observations on exit from the motor insurance industry, 1966–1972' *Journal of Industrial Economics*, June 1974.

29 Westall, 'Invisible hand', pp. 444–5.

30 R. L. Carter, *Competition in the British Fire and Accident Insurance Market*, D.Phil. thesis, University of Sussex, 1968.

31 Beale, *V&G Crash*, chapter 4; *V&G Inquiry*, chapters III and IV.

32 Carter, *Competition*, p. 76; on the inflexibility of the tariff see the *Economist*, 13 July 1968, Insurance Supplement, pp. ix–x.

33 On McKinsey see the *Economist*, 14 May 1966, p. 727 and 23 July 1966; *PM*, 13 May 1965, p. 761, 20 May 1965, p. 781 and 7 July 1966, p. 1119.

34 *PM*, 21 July 1966, p. 1213 and 28 July 1966, p. 1249.

35 The *Vehicle and General's* accounts are summarised in Appendix D of the *V&G Inquiry*; Lloyd's domestic motor income was published in *PM*, passim.

36 Beale, *V&G Crash*, chapter 6.

37 *PM*, 24 October 1968, p. 1969; the *Economist*, 12 October 1968, pp. 71–2 provides an excellent analysis of the collapse of the motor tariff.

38 *PM*, 5 February 1970, p. 259; Beale, *V&G Crash*, pp. 67–73.

39 *PM*, 17 June 1971, p. 1179

40 Beale, *V&G Crash*, chapter 6.

41 *Ibid.*, chapter 8; *V&G Inquiry*, chapter V and Appendix D; Franklin, 'Exit from motor insurance', p. 305.

42 *PM*, 11 March 1971, p. 483.

43 *V&G Inquiry*, chapter x; Tapp, 'Regulation', pp. 34–5.

44 Tapp, 'Regulation', pp. 35–6; Peter J. Franklin and Caroline Woodhead, *The UK Life Assurance Industry: A Study in Applied Economics*, London, 1980, pp. 345–53.

45 *Ibid.*, p. 349.

46 *Ibid.*, pp. 354–6.

47 Franklin and Woodhead, *UK Life Assurance Industry*, p. 345.

48 'As time has passed, the elaborate tariff structure of agreements and committees has come increasingly to resemble the Maginot line — formidable but inflexible and progressively more irrelevant to the real problems.' *Economist*, 13 July 1968, Insurance Supplement, p. ix; see also Carter, *Competition*, pp. 80–1; there were several 'insiders' criticising the tariff system in the late 1950s: see *PM*, 15 August 1957 and 21 August 1958.

49 Monopolies Commission, *Fire Insurance*, pp. 4–6 and chapter 3 provide information on independent fire insurers.

50 Monopolies Commission, *Fire Insurance*, p. 38.

51 Carter, *Competition*, p. 221–2.

52 *Ibid.*, p. 87.

53 The BIA loss estimates were published in *PM*, passim.

54 *PM*, 9 February 1961, p. 183; 10 December 1964, p. 1851; and 12 June 1969, p. 1115.

55 Carter, *Competition*, p. 108.

56 *PM*, 4 July 1963, p. 949; Monopolies Commission, *Fire Insurance*, pp. 41–2.

57 *PM*, 6 August 1964, p. 1195.

58 *PM*, 16 April 1964, p. 579.

59 *PM*, 21 January 1965, p. 99

60 G. C. Allen, *Monopolies and Restrictive Practices*, London, 1968, p. 126.

61 *Monopolies, Mergers and Restrictive Practices*, March 1964, Cmnd. 2299.

62 Monopolies Commission, *Fire Insurance*, p. 6.

63 *PM*, 5 March 1970, p. 465 and 21 January 1971, p. 145; Monopolies Commission, *Fire Insurance*, p. 21.

64 Monopolies Commission, *Fire Insurance*, chapter 7.

65 *Ibid.*, chapter 8.

66 *Economist*, 19 August 1972, p. 76.

67 *DTI Press Release*, 19 April 1985.

68 *Economist*, 9 June 1973, Insurance Supplement.

69 Carter, *Competition*, pp. 39–41 and 212–16; R. L. Carter, *Economics and Insurance*, Stockport, undated, p. 9.

70 Information supplied by DTI.

71 DTI Press Release, 19 April 1985.

72 Porter, *Competitive Advantage*, especially chapters 6 and 12.

9 *T. A. B. Corley*

Oil companies and the role of government: the case of Britain, 1900–75

1 The issues

Competitiveness can be defined in two ways. At the macro-economic level, a country will be competitive if its average unit costs are below those of its rivals, while at the industry level, the degree of competitiveness will depend on such factors as low or high barriers to entry. In devising any policy of encouraging competitiveness, therefore, the state faces a dilemma. On the one side, it may plan to control mergers or monopolistic practices which allow firms to force up prices and to permit X-inefficiency. On the other side, in the present-day world dominated by large companies, it might choose actively to assist its own giants as promoters of technological and organisational innovations which serve to keep an economy ahead of its rivals.

During the present century, British governments have repeatedly encountered this particular dilemma in their dealings with the oil industry. Largely for commercial reasons, leading oil companies have to be of considerable size, and are usually integrated, in being concerned with production, refining and the marketing and distribution of refined products. Yet over the years there has been regular criticism by consumers and political or pressure groups in Britain that the oil giants have been overcharging the public. The present account will make clear that while successive governments have not pursued a consistent strategy over oil, they have had to operate within certain fundamental constraints which broadly guided their decisions.

The main constraint was that, until North Sea oil came on stream in the 1970s, Britain lacked adequate indigenous sources of oil and had to rely on imports. From the nineteenth century onwards, Scottish shale oil enjoyed some prominence; however, after 1918 the industry was languishing and expired in the 1960s, while a few inshore mineral oil wells made an insignificant contribution to supplies. By contrast oil's principal substitute, coal, had for centuries provided heat and indirectly light and power for firms and households. Although petrol had to be used for motor transport once horse-drawn vehicles were phased out, for other users oil had the advantage of being far more convenient to handle and of relatively higher thermal content. Yet for both strategic and political reasons the British coal industry could not be allowed to expire, and government found itself walking a tightrope in seeking to balance the demands of the relative interest groups.

However, overall policy was concerned with ensuring adequate supplies from overseas, even at considerable balance of payments cost. The authorities' broad success here, despite restrictions during two world wars and later problems caused by crises in the Middle East, made sure that Britain's ability to compete was not to any great extent harmed by a shortage of oil products as the economy became increasingly oil-dependent. At the same time, governments were normally unwilling to promote national competitiveness by forcing the oil giants to hold down prices. Fuel costs were after all a not insignificant element in industry and commerce, from energy in factories to the transport of raw materials and finished goods and the forcing of horticultural products in these cold and damp islands.

The narrative, in the section below, of relations between the state and oil companies, will therefore be within the context of a long-standing government preoccupation with supplies but a willingness to make users pay at a rate that was high by international standards.

2 State–corporate relations in Britain, 1900–75

2.1 *The background*

The first supplier of mineral oil products had been America, despatching kerosene to Britain from 1861 onwards, two years after Colonel Drake's initial oil strike there. Russia then became a significant supplier in the 1880s. By 1900 kerosene comprised 84 per cent of Britain's oil imports and lubricating oil 16 per cent. There was a free market, with British government intervention restricted to laying down safety regulations for the handling and storage of these potentially hazardous products. Despite the downstream monopolistic activities of Rockefeller's Standard Oil Company in the US, competition allowed unit prices to fall by 40 per cent between 1885 and 1914. Britain's imports were in the form of refined products, and until petrol became important after the turn of the century, the overseas refineries largely burnt off the lighter fractions, such as naphtha, as commercially valueless.[2]

As the account given below will reveal, government intervention was to fluctuate despite the acceleration in the pace of change during the twentieth century. Table 9.1 distinguishes five periods with varying degrees of involvement between 1900 and 1975. The successive sub-periods will then be examined in turn.

Table 9.1 *Phases of state–corporate involvement in oil, 1900–75*

Period	*Degree of involvement*	*Main determinants*
1900–25	Relatively high	Need for reliable source of fuel oil for warships 1914+. First World War and oil control
1925–39	Relatively low	World recession. Oil glut — successive cartel arrangements to maintain 'orderly market'
1939–54	High	1939+. Second World War Post-1945. Dollar shortage — need for home refining 1951–4. Dispute with Iran — long-term global repercussions
1954–70	Reduced	Rapid growth in low-cost sources of oil world-wide. Consequent glut of cheap oil diminishes need for state intervention. North Sea oil — state licensing procedure required at end of period
1970–75	High	Disruptions caused by OPEC activities. Embargoes and price rises

Source: Derived from Louis Turner, *Oil Companies in the International System*, London, 1978.

2.2 *The emerging role of oil, 1900–25*

Whereas between 1861 and 1900 Britain's oil needs had been furnished through a developing market without any noticeable supply or price problems, the next quarter of a century saw a growing government concern over these problems as the role of oil in the economy expanded significantly. Oil transformed itself from being principally an illuminant into a highly versatile source of energy. Hitherto undervalued by-products thereupon became important in their own right; by 1913 petrol and fuel oil made up 40 per cent of Britain's oil imports. How, and why, was government drawn into oil affairs?

The market continued to govern supply to the consumer of kerosene and petrol, with excessive safety regulations holding back the growth of the latter's bulk distribution. At the same time, genuine competition among the three main petrol supplying companies was hampered by a market-sharing agreement and the hardening of prices from 1912 onwards led to a public outcry, including demands for government curbs. Those were not forthcoming over petrol; instead, it was fuel oil which engaged the attention of the Admiralty, which was striving to substitute oil for coal as fuel in British warships, so as to permit a considerably greater radius of action, opportunities for higher speeds and more efficient use of personnel afloat. Geoffrey Jones's pioneering studies have shown how the need to implement that substitution proved to be a landmark in the emergence of an oil industry in Britain.[3] Yet for a whole decade the Treasury consistently declined to help British oil magnates who sought funds in return for long-term fuel oil contracts, from Austen Chamberlain's refusal in 1903 to lend William K. D'Arcy £300,000 on the Persian concessions to the rejection in 1913 by a later Chancellor, Lloyd George, of Lord Cowdray's request for a £5 million investment in the latter's Mexican Eagle Oil Company.

A key pointer to closer official involvement, not mentioned by Jones, was the interest taken in oil matters by the Committee of Imperial Defence, the top executive body responsible for strategic planning throughout the British empire from 1905 onwards. The only part of the empire with significant proved oil deposits was Burma, a province of India. The government of India had decided, on competition grounds, to restrict the number of concessions granted to the largest producer, the Burmah Oil Company. However, when that company's existence was threatened by foreign rivals early in 1905, the Committee circularised all interested government departments about the 'vast importance' to British interests of the Burmese oilfields, and pressed for Burmah Oil's directors and shareholding to be made exclusively British. Although such steps were never taken, later that year the Admiralty signed with Burmah Oil a contract to supply fuel oil from Rangoon, on a limited scale in peacetime but very substantially in the event of war.[4]

As it happened, only modest quantities were taken up since Rangoon was too remote from the main fleets in home waters and the Mediterranean. Even so, the Admiralty valued the contract as a bargaining counter against overcharging by established suppliers, notably Shell and Standard Oil. The goodwill which the Burmah Oil directors built up with the Admiralty led to their being offered the chance to acquire D'Arcy's oil concession in Persia when he was forced to sell, so as to prevent it from being lost to Britain altogether. They took over operations there, and in 1908 their drillers struck oil. A year later, they set up the Anglo-Persian Oil Company — subsequently British Petroleum — as a subsidiary of Burmah Oil. It was Winston Churchill, First Lord of the Admiralty from 1911 onwards, who induced the Treasury in 1914 to purchase a majority shareholding in Anglo-Persian, and in return obtained a long-term fuel oil contract. As Persia was in the British sphere of influence, that represented a virtually secure source.

The First World War From 1914 to 1916, Whitehall ran the war on a 'business mainly as usual' basis, relying on the market for most resources needed, including oil. The petrol tax was doubled and prices rose by about a third, but supplies were plentiful enough for rationing to be unnecessary. Since Anglo-Persian was not yet technically able to provide fuel oil, the Admiralty had to buy most requirements from Standard Oil and Shell. Another of its suppliers was Mexican Eagle; yet relations with Lord

Cowdray were not easy. American oil interests were hostile to Cowdray, who they believed was being unduly favoured by the regime in Mexico, and he felt let down by the Foreign Office. The latter placed the aim of fostering harmonious Anglo-American relations above actively supporting him as a British subject.

Not until early 1916, with sharply declining imports of oil products, did government departments intervene in the market, for the first time drawing up estimates of demand and supply and designating the Board of Trade to administer oil affairs. Petrol was at last rationed, and successive restrictions eventually ended private motoring for non-essential purposes. At the same time, a number of overlapping official committees were established, and in 1917 a body of oil men, the Petroleum Pool Board, was set up to regulate in detail the increasingly scarce supplies. Companies were allotted fixed percentages, according to their previous market shares, and were granted sole distribution rights in specific zones. To prevent profiteering, the government controlled both prices and distributors' profit margins. By the end of 1917 a Petroleum Executive came into being to confer systematically, rather than on the earlier *ad hoc* basis, with other departments and devise an overall strategy for oil.[5]

The merger scheme Wartime emergencies forced the government to consider longer-term remedies for Britain's strategic weakness over oil. One possible solution was to create an all-British oil combine. In 1915 the chairmen of Anglo-Persian and Mexican Eagle proposed a merger of their companies: that foundered partly because Whitehall feared that the US administration would retaliate by barring UK oil ventures there if Mexican Eagle were to become formally linked with the government-controlled Anglo-Persian. In 1919 Cowdray, having suffered further examples of Whitehall's indifference to him, sold his Mexican interests to Shell.

The second merger plan was to amalgamate Burmah Oil, and later a very reluctant Anglo-Persian, with Shell's British assets into a similar combine. The Foreign Office in particular, aware that Britain was relying heavily for her oil needs on the Shell group, which was 60 per cent Dutch-owned, pressed hard for that group to be brought under British control. Yet neither scheme achieved fruition, being unable to guarantee entirely secure supplies of oil. It was more important for a proportion to come from a reliable source, such as Persia, to provide an element of competition.

Post-1918 Britain's first 'total' war amply demonstrated the key role played by oil products, with aircraft and land vehicles, including tanks, as well as warships and merchant vessels, running on oil. Yet the government refused to adopt dirigiste policies towards oil, unlike the French, who set up the semi-public Compagnie Française des Pétroles and insisted on domestic refining. On the contrary, it worked towards disengagement. Although

Britain benefited from Anglo-Persian's pursuit of concessions globally and other measures to turn itself into an integrated giant of world stature, the Treasury refused to provide more capital for ventures outside Persia. Conservative ministers were even willing to privatise Anglo-Persian, as part of the second merger scheme, which was not finally abandoned until the MacDonald government of 1924 bowed to the widespread public hostility towards the oil trusts. Similarly, the Foreign Office was prepared to sacrifice Britain's global oil interests to pressures from Washington, by then gravely concerned about US oil shortages, so that American companies ended up with stakes in Mesopotamia, later Iraq, and subsequently in Saudi Arabia, Bahrain and Kuwait.

After wartime controls were wound down in 1919, Whitehall did nothing to curb the power of the established suppliers. An official committee on prices and profiteering issued two reports in 1920–21 on petrol, finding prices and costs to be excessive.[6] It suggested that prices should be fixed under the Profiteering Act and that the League of Nations should sponsor international protection against exploitation by the oil trusts, as well as active research into possible substitutes for petrol. Britain took no action on these reports, and the oil companies maintained their wartime co-operation and agreed selling prices among themselves, competing only on the services offered.

One positive move was that the Petroleum Executive was in 1920 made into a department of the Board of Trade. Comprising a small nucleus of civil servants with an informed knowledge of oil affairs, it advised ministers and officials of other departments who knew little or nothing about oil. Its technical expertise was particularly helpful during the complex oil negotiations over the Middle East, and helped to limit the harm done by excessive Foreign Office subservience to the US and other countries. However, by 1934 the Petroleum Department had been reduced to no more than three officials working full time and a further three part time.[7]

2.3 *Retreat by government, 1925–39*

The abrupt ending of the post-war boom caused oil prices to decline, but even so, by the mid-1920s supply exceeded demand. The world's oil giants therefore took joint steps to maintain an orderly market through restrictive agreements. In 1927, for instance, Anglo-Persian and Shell set up a joint marketing organisation for much of the eastern hemisphere. A year later these companies, and Standard Oil of New Jersey, signed the celebrated Achnacarry agreement, which regulated current and prospective market shares virtually world-wide. They aimed both to forestall damaging price wars and to minimise distribution costs by drawing supplies from sources nearest to customers.[8] At the same time, they imposed a unified pricing system based on the Gulf of Mexico, which allowed product prices in many parts of the world to be higher than they needed to be.

Preparations for war Britain's Committee of Imperial Defence maintained its responsibility for overall supply planning in case of war. In 1952 it set up a sub-committee, the Oil (Fuel) Board, under ministerial direction, for programming both oil and tanker requirements. Yet the cabinet's ten-year rule of 1919, that the armed forces should plan on the assumption that no major war would occur for ten years, persisted until the early 1930s. Not until 1934 was the Oil Board able to prepare actively for possible wars in Europe and the Far East. From 1936 onwards, as soon as a general conflict became likely, the Petroleum Department's inadequate staff was built up again.[9]

2.4 *The Second World War and its aftermath, 1939–54*

Wartime control[10] In Britain, Second World War planning was put in hand from the outset, and the Petroleum Department had the task of relating military and civilian needs to the country's overall war effort. Detailed allocation plans were then left to the industry-manned Petroleum Board, chaired by a top oil figure. All firms pooled their supplies and were granted quotas. Petrol was rationed and prices were strictly controlled. The consequent informal relationship between civil servants and oil managers has persisted until the present day.

In 1942 the Petroleum Department's role was strengthened by its becoming a division within the newly created Ministry of Fuel and Power. After America's entry into the war, joint committees for oil were set up. Although the US administration pressed for rigid overall control by the military, Britain won the principle of maintaining an essentially civilian and semi-formal system of liaison between officialdom and the oil industry. Hence no Combined (UK–US) Oil Board was ever established, but a Petroleum Mission in Washington kept the administration informed of events in London.

Even so, US officials mistrusted Britain on two grounds. First, Britain was believed to be selfishly holding on to huge Middle Eastern oil reserves just when America was again running short of oil. In fact, the paramount need to save tanker capacity dictated that the maximum use should be made of western hemisphere oil in the war against Germany. Second, Washington was suspicious of British oil companies' cartelisation, as the US Federal Trade Commission's influential report of 1952 was later to reveal. The repercussions of such mistrust were to surface from time to time in the future.

Post-war measures in Britain Wartime controls over oil were relaxed only slowly after 1945, petrol being rationed until 1950 and price controls not being abolished until 1953. With the abrupt ending of lend-lease, Whitehall's concern at the dollar shortage intensified, as Britain's and the empire's oil requirements came largely from the western hemisphere. Crude oil was still refined near to the point of production, and as late as 1939 no less than 80 per cent of imported oil reached Britain as refined products.

Then in 1947 Hugh Gaitskell, newly appointed Minister of Fuel and Power, began to persuade the companies to refine in Britain. That would reduce the dollar burden of oil and later on allow them to use the cheaper crude oil from the Middle East as it became increasingly available. He also hoped to encourage a petrochemical industry, as well as domestic manufacturing facilities for refining equipment, hitherto obtainable only from the US. By 1960 about 77 per cent of Britain's oil needs were imported in crude form.

In the 1950s, demand for oil products of all kinds soared, but by then could be paid for largely in soft currencies; Iraq was in the sterling area until 1958. The foreign currency burden of imports was partly offset by the overseas incomes earned by Anglo-Iranian (as Anglo-Persian had become in 1935) and Shell, which both operated in sterling.

The Abadan Affair, 1951–54[11] The crisis precipitated in 1951 by Dr Mossadeq's nationalisation of the Anglo-Iranian assets in Iran is significant for two reasons. For the last time the British government reacted to an oil crisis virtually on its own, and the attitude of the US authorities influenced Britain's strategy even more than in earlier decades. The US authorities initially supported Mossadeq as a bastion against communism, and not until Eisenhower became president in 1953 did American policy change. That August an undercover operation by MI6, inexpertly assisted by the CIA, engineered a coup which established a more pro-Western regime in Tehran. An agreement in 195 left Iran in control of upstream operations, while an international oil consortium took over marketing, Anglo-Iranian's interest being no more than 40 per cent. That company, renamed British Petroleum (BP), rapidly undertook a further diversification of its activities, and forged ahead in the rest of the world.

2.5 A laissez-faire *interval, 1954–70*

In 1950 oil accounted for no more than 10 per cent of Britain's energy requirements, but by 1973 that share had risen to over 48 per cent. Meanwhile, oil production in the non-communist world was growing annually by over 7 per cent between 1957 and 1970, with proved oil reserves showing a comparable increase.[12] The new fields were largely in the Middle East and North Africa, producing at exceptionally low unit cost.

Thus during this period the market was able to satisfy British oil needs ever more cheaply in real terms. Yet the government and the retailers benefited more than the final users. Successive chancellors of the exchequer regularly increased petrol tax rates and from 1961 onwards placed a duty on heating and other heavy oils, thus benefiting general revenue. Whitehall accepted relatively high oil prices because of its need to support the indigenous coal industry — hence the tax on competing heavy oil — and an ambitious nuclear energy programme. By the mid-1960s the UK's pre-tax petrol prices were 10 per cent higher than in Italy, 18 per cent than in France and 21 per cent than in Switzerland.[13] Yet the British economy was badly in need of low oil prices as global competition intensified. Germany and Japan were rejoining world markets, the EEC was turning itself into a formidable economic bloc, and tariffs were being successively reduced.

Although, as Helen Mercer has shown in Chapter 5, the Monopolies Commission had been operating since 1949 and repeated parliamentary questions had pressed for the investigation of petrol prices, the Foreign Office and Minstry of Fuel and Power successfully lobbied the Board of Trade to save the activity of petrol distribution from referral. Not until 1965 did the Commission, after five years' study, issue a report on that topic. As it admitted candidly, 'the history of petrol supply in this country shows that since the invention of the motor car the major suppliers have consistently (except when matters were taken out of their hands in time of war) sought by one means or other to minimise price competition'.

The Commission therefore concentrated on the solus system, whereby most UK petrol stations were tied to a single oil company, receiving rebates unrelated to the quantity delivered at any one time. The report did not condemn the solus system outright, but showed retailers' profits to be unnecessarily high, partly through a distribution and pricing system which — for instance in maintaining price zones — was largely unchanged from wartime control days. Here could be seen the arthritic condition of an important and expanding sector of Britain's economic life. The government failed to insist on a radical overhaul of the system, but simply barred the large companies from further extending their networks of retail outlets. Even that restriction was abolished in 1968, when the companies undertook to increase their refining capacity at home.[14]

Suez and after, 1956 The Suez crisis effectively began and ended with oil: soon after Nasser's seizure of the canal, the Minister of Fuel and Power warned the British cabinet of the risk to western Europe's oil supplies, at a time when Britain held no more than six weeks' stocks. Ultimately, the Anglo-French invasion of Egypt was halted as much by American threats to cut off oil as by the concurrent run on sterling. To conserve British supplies, a huge but temporary surcharge was imposed on petrol, which was also rationed.

Whitehall learnt the lesson of having been caught with inadequate oil reserves, and persuaded the oil companies to build up stocks. The Organisation for European Economic Co-operation — in 1961 broadened out into the OECD — went further and set up a Petroleum Energy Group, for the discussion of common problems with the principal companies involved, a step rightly hailed as breaking new ground in international collaboration.[15] It also devised an allocation scheme for member countries in the event of further crises, and sought to make higher stocks mandatory on all members. Hence the OECD weathered the Arab-Israeli war of 1967, a further closure of the Suez canal and Saudi Arabia's oil embargo against Britain and the US. At the same time, the administration's anti-trust laws debarred American oil corporations from working systematically with the Energy Group, or even consulting on a joint basis with the State Department.

Developments in Whitehall The post-war expansion in the role of the British state seems to have brought about a division between two kinds of ministry: the 'broad–brush' departments and those concerned with more detailed and routine affairs. The former were usually small in size, and basically engaged in tackling large issues involving general principles. Headed mainly by Oxbridge firsts in the humanities, they would strive to reflect the political views of their ministerial chiefs. The latter tended to be newer creations to administer often complex acts of parliament and delegated legislation. Their mandarins, once they had gained the somewhat narrow expertise required, might be tempted to 'go native' and perhaps over-identify with their clientele, whether individual or corporate.[16]

Sometimes broad–brush departments were given specific tasks of the latter kind, and the civil servants charged with these tasks apparently became soft towards the public they served. The Board of Trade (Chapter 5), which from 1948 onwards oversaw the Monopolies and Restrictive Practices Act, to a certain extent found itself 'captured' by the business interests it was supposed to keep in check. The Ministry of Agriculture (Chapter 10), under the Agriculture Act of 1947, administered the system of price support and deficiency payments. While the hostility between minister and the vociferous farming lobby persisted, there were recurrent complaints that officials were far too close to the large chemical and food processing companies.

In the case of oil, the Foreign Office can be contrasted with the Ministry of Power, as it was renamed in 1961. Foreign Office policy was strongly swayed by the need to placate the US administration, and that need grew more urgent as Britain's economic weakness and balance of payments problems intensified after 1945. On the other hand, the petroleum division of the (Fuel and) Power ministry was very close to the oil companies, receiving confidential information which the civil servants were not inclined to pass on to their notoriously leaky masters. These conflicts of view between generalist and specialist departments erupted in the two oil episodes to be considered next, those of the North Sea discoveries and Rhodesian sanctions.

North Sea oil[17] Britain's competitive weakness, owing to the virtual absence of indigenous oil and her consequent dependence on overseas supplies, was pointed out earlier. That source of weakness unexpectedly if temporarily vanished when offshore oil was discovered in the North Sea. The government lost no time in taking the initiative, to guide and control the exploitation of this huge windfall of resources and thereby act as a market completer.

In earlier decades, UK private industry had been incapable of developing for peaceful purposes that other source of power, nuclear energy, so that the public sector had had to take over that task. Oil companies, on the other hand, possessed both the financial muscle and the know-how to realise the opportunities in the North Sea, in part thanks to decades of earlier offshore ventures in many parts of the world. Given these favourable circumstances, co-operation between Whitehall and the oil companies remained good. Once a United Nations convention in 1958 had granted Britain rights over the extensive continental shelf of the North Sea, the companies gladly provided officials with advice on the technicalities involved in drafting the appropriate legislation. In 1964 the government ratified the convention and passed into law the Continental Shelf Act.

Although the Foreign Office would have liked Britain to offer EEC countries some of her oil as a sop to help win membership, the North Sea sector involved stayed in British hands and the Ministry of Power became the allocating body. Rather than auctioning the whole or parts, it preferred to issue licences on a discretionary principle, which permitted closer control as well as preferential treatment to British companies or consortia. It divided up the whole sector into 100 square-mile blocks, and required of successful applicants proof that they had the necessary expertise and could make a material contribution to North Sea development. To help them to get ahead of their foreign rivals, BP and Shell were granted stakes in the most favourable areas, soon gaining proved reserves even greater than their participating shares, which rose from 30 to 43 per cent in the licensing rounds to 1971.

Despite this sensible licensing policy, the government's overall North Sea strategy was criticised on three counts. First, it was not sufficiently tough over fixing the boundary line with Norway to the east. Informed opinion held that if Britain had been willing to take Norway to the International Court of Justice at The Hague, she could have secured far more than the barely generous 35 per cent of the North Sea with which she ended up. The hand of the Foreign Office can be seen behind that craven decision. Second, the government was too lenient to the companies. Third, it did little in the way of encouraging British firms to manufacture the ancillary equipment, such as rigs and platforms, to service the North Sea bonanza. No doubt any pressure by the Ministry of Power on the Board of Trade and other departments proved ineffective. In any case, the former's influence in

Whitehall dwindled when in 1969 it was absorbed into the Ministry of Technology. A giant Department of Trade and Industry, set up by the new government of 1970, took over oil affairs, and in line with Conservative policy made clear that the ability to supply North Sea equipment was a matter for the industries themselves.

The second criticism came to a head in 1973 when the House of Commons Public Accounts Committee, investigating North Sea oil and gas, revealed that participating oil companies were grossly undertaxed, having contributed less than £500,000 in all as tax payments on UK operations.[18] Indeed, the Exchequer was receiving a far lower share of oil revenues than in any other comparable country. Since 1965 British corporation tax had been assessed on profits world-wide, with companies being able to offset their fiscal obligations against exploration outlays both at home and overseas. The corrective measures taken by Labour after 1974 will be discussed below.

Rhodesian sanctions, 1965–70 This episode caused the most serious rifts ever between government and the oil companies. After the white regime in Rhodesia proclaimed UDI in 1965, the British government imposed oil sanctions in accordance with a United Nations resolution. However, the precedents for successful enforcement in peacetime were not encouraging. In 1935 Italy had invaded Abyssinia, a member of the League of Nations, whose council at once declared financial and other sanctions against the aggressor, with oil to be added to the list as soon as the leading powers gave their approval. Then France took fright when Mussolini threatened that sanctions would mean war.

A divided British cabinet, unwilling to authorise sanctions itself, instructed the Board of Trade to ask Shell and (the then) Anglo-Iranian to deny oil to Italy on their own initiative. The companies knew that, in the absence of *force majeure*, they would become liable to substantial damages if they repudiated contracts, and therefore refused. The Petroleum Department supported the companies and warned the cabinet that sanctions could never be made watertight without the co-operation of US companies, a condition ruled out through fear of anti-trust reprisals from Washington. Early in 1936 a still disunited cabinet reluctantly agreed to sanctions, but having failed to secure backing from the other powers, it then had to let the proposal lapse.[19]

In 1965 its successor was unanimous in approving a sanctions order against Rhodesia. Yet the problem of securing effective international co-operation remained. Washington had approved sanctions only after Britain had promised as a very costly *quid pro quo* to maintain her armed forces in the Far East. Even so, the Americans were privately sceptical about sanctions working because Rhodesia's neighbours, South Africa and the Portuguese colony of Mozambique, supported the white cause. As will be seen, neither France nor Portugal would fully back Britain. Consequently ministers were forced to retreat from an overweening confidence that resolute measures would swiftly bring down the rebel regime, to an anxiety that no UK oil should be seen to be getting through, and finally to a recognition that British oil companies were at least indirectly breaching sanctions orders but with a pious hope that they would never be found out. The sanctions fiasco was so complete that in 1969 Rhodesia consumed as much oil annually as before UDI and by 1978 had doubled its total consumption.[20]

Britain's further difficulty here was that by the 1960s oil majors were operating world-wide and had in consequence so decentralised their structures that branches automatically identified themselves with the interests of the host countries concerned. As the Foreign Office and Treasury were not prepared to see oil sanctions extended to South Africa and thereby put British investments there at risk, much oil was likely to enter Rhodesia from that source. Mozambique, too, was known to be allowing supplies through to Rhodesia, but ministerial pleas with Portugal were brushed off in Lisbon. The Royal Navy did impose a sea blockade on the Mozambique port of Beira and continued that futile and expensive operation until the colony became independent in 1975. When France was shown to be clearly implicated in sanctions busting, Harold Wilson requested the President, General de Gaulle, to halt the practice, but de Gaulle refused.

Foreign Office diplomacy was therefore proving impotent as a means of enforcing sanctions. Admittedly, neither that department, nor the Commonwealth Office which controlled government sanctions policy until amalgamated with the Foreign Office in 1968, had much enthusiasm for sanctions, given the tepid attitude of the US and a fear of stirring up real trouble in France, Portugal, South Africa and elsewhere. The half-heartedness of Ministry of Power officials sprang from a different perspective: that, given current political and other realities, the oil companies involved would find it difficult if not impossible to deliver watertight sanctions. Those officials' pragmatic stand made them unpopular among politicians, who were convinced of their subservience to the oil lobby. Harold Wilson complained that one of his ministers of power had been 'flatly refused' any statistics by the oil companies, while the Treasury, the responsible department for BP as holder of a majority of its shares, was unable to extract vital information from that or any other company.

The sanctions débâcle unfolded from 1968 onwards. After several years of seemingly close contact with the Commonwealth Office, that year BP and Shell owned up to officials that some of their oil was ending up in Rhodesia. They informed the responsible minister to that effect, claiming that they had 'at once done what they safely could do to put things right'. Harold Wilson later claimed that he did not know of this admission. Controversially, the companies then arranged to swap oil in South Africa with the French company Total, which kept Mozambique, and hence indirectly Rhodesia, supplied. Although they fully briefed Ministry of Power officials — who kept their ministers in ignorance — no more than hints were passed on to the Commonwealth Office. Hence for those in government and in the oil companies, concealing uncomfortable facts was more important than striving to make sanctions work.

Conclusion To sum up on this period, by the end of the 1960s the oil companies had revealed what could only be termed self-serving attitudes over both the North Sea and the Rhodesian sanctions episodes. The companies clearly needed to be kept on a far tighter rein by government. It will be seen below how an even graver crisis precipitated UK government intervention that was more sustained than at any other time except in total war.

2.6 *The onset of OPEC, 1970–75*

By the late 1960s consuming governments were placing their trust in an oil market that they hoped would be capable of adjusting itself within about two months to any upsets in supply: the EEC had followed the OECD's lead in requiring members to hold stocks equivalent to sixty-five days' use. Companies, too, relied overmuch on their considerable world-wide strength to face down the Middle Eastern producing countries, which since 1960 had come together in OPEC. It was in fact market forces which brought about the crisis of 1973–74; the Yom Kippur war of 1973 was simply the occasion.

Causes of OPEC crisis[21] The huge world surplus up to 1966, which had allowed companies to market oil so cheaply, thereafter fell from seven million barrels a day in 1966 to below two million in 1973, with the largest consuming country, the US, finding its surplus cut virtually to zero by 1973. To compound their difficulties, advanced countries could no longer rely heavily on alternative energy sources. Most had drastically reduced coal production, while the glittering reputation of nuclear power was already becoming tarnished. When in 1971 two OPEC members, Libya and Iran, took advantage of the emerging market disequilibrium to extract stringent terms from their respective oil companies, neither consumer governments nor companies were capable of organising concerted resistance to the growing assertiveness of OPEC.

The reactions of the British government The precise response of Whitehall to these far-reaching developments will not be known until the public records are open after 30 years. Yet apparently in the early months of 1972 officials in the petroleum division of the Trade and Industry department faced up to the difficulties ahead. Britain's largest single oil supplier, Saudi Arabia, providing a fifth of total UK requirements, had recently demanded greater participation from the US-owned Aramco. In the face of the latter's obduracy, the Saudis looked round for an oil company not directly involved in the Middle East, which could be persuaded to take over the marketing of their crude behind Aramco's back. With the assistance of Whitehall, they approached Burmah Oil, then short of crude for operations at its Bahamas terminal, where oil was transhipped to east coast ports in the US which were too shallow to take supertankers. Aramco eventually climbed down, but Burmah Oil continued talks with Saudi Arabia until they were overtaken by the war of 1973 and its consequences.[22]

At that crucial time, Whitehall was paying a heavy penalty for having absorbed the Ministry of Power into the overlarge Department of Trade and Industry. The latter was known to be quarrelling with the Treasury and the Foreign Office as to which department ran oil, so that there was no overall oil strategy. The Conservative government ostensibly adopted a hands-off attitude to oil questions, expecting the companies themselves to carry out negotiations but standing ready to assist if requested.[23] Not until December 1972 did the EEC agree to raising stocks to 90 days' use, even then not to become fully effective until 1975. In mid-1973 EEC members did strive to hammer out a common energy policy, only to have their efforts thwarted by France which resented Britain's still substantial influence in world oil affairs. Fortunately for Britain, while French and American oil companies were at loggerheads with one another, BP and Shell were working reasonably well together. BP's huge venture in Alaska presented no conflict of interest to Shell, and the two regularly swapped technical information about the North Sea operations.

The Yom Kippur war and its consequences, 1973–74[24] In conditions of severe strategic weakness, consumer governments and companies were unprepared when the outbreak of the Arab-Israeli war in October 1973 was followed by OPEC's severe cutbacks and also embargoes against certain consumer countries. Until the archives are open, British government response will be only partially known, but available evidence suggests that the broad-brush departments were not adequately briefing ministers, especially the Prime Minister, Edward Heath. While reassuring parliament that Britain had seventy-nine days' stocks and therefore did not need to ration petrol, he was privately exerting pressure on the chairmen of BP, Royal Dutch and Shell. He demanded preferential treatment for Britain, but was told that the companies must treat all customers equally and ration supplies across the board in line with Arab cutbacks. Otherwise they would incur heavy damages were they to break contracts arbitrarily.

Heath claimed that Britain, like France, had special treaties with Middle Eastern producing countries which would ensure her normal allocations of oil, but when asked to produce the texts he was unable to do so. On his arguing that BP was half owned by the government, he was informed that the share agreement of 1914 granted the company full operational freedom, and that the honouring of contracts came before shareholders' interests. The chairmen could only suggest that parliament might pass a law establishing *force majeure*; in that case they would require from him a list of countries to be penalised so as to maintain supplies to Britain. Heath declined both suggestions.[25] Instead, the Foreign Office promoted some diplomatic missions to Middle Eastern countries, with the hope of concluding bilateral agreements for the exchange of oil for industrial products and in particular armaments. These initiatives faded away once the cutbacks and embargoes were relaxed.

Petroleum division officials were clearly more skilful in damage limitation, and successfully finessed Exxon into stepping up crude shipments from company sources to keep open its recently expanded refinery in Britain. They hinted that Exxon would wish to protect its 25 per cent share of the British oil market. Meanwhile, the oil companies raised the price of oil products to consumers by more than the OPEC increases, thereby helping to restore their profit margins which had been eroded during the lengthy period of low prices. Not until 1976 did the Monopolies Commission again look at the petrol distribution system in Britain, reporting in 1979.[26]

The oil regime in the UK from 1974 onwards The oil crisis did persuade Heath to establish a Department of Energy in January 1974, oil policy within that department being sensibly split between a home and an overseas division. An urgent need was to act on the pungent criticisms of the recently issued Public Accounts Committee report. The incoming Labour government in March at once took steps to deal with the two main criticisms. To make up the revenue shortfall from the oil companies, it introduced a petroleum revenue tax and designated each field as a separate tax point to halt the earlier avoidance measures. It also set up a wholly state-owned and integrated British National Oil Corporation (BNOC), along the lines of Norway's Statoil. That would manage the 51 per cent interest in all North Sea concessions which the government planned to acquire, and would compete directly with the private sector, ensuring that the rate of depletion was properly controlled and that the oil would be sold to the British economy's maximum advantage. On the second criticism, about neglect of the opportunities created for British industry generally, the government as a large purchaser sought to favour contractors at home and pressure all parties concerned into 'buying British'.

To provide BNOC with the required nucleus of integrated activities, the government expected BP to hand over all its resources in the North Sea. BP, having seen off Edward Heath, was no more amenable to coercion from his successor, and refused to deliver.[27] In 1976, therefore, BNOC had to be set up with no co-operation from its rivals, which resented this newcomer as an agent of the state with favoured access to capital funding. By an unexpected windfall, the Department of Energy was able to secure the North Sea exploration staff of the then stricken Burmah Oil Company, in return for a favourable deal over part of that company's North Sea assets.

BP's independent stance highlighted its anomalous position in an era of enhanced government intervention. It needed to be either wholly in the public or the private sector. Tony Benn as Industry Secretary and then as Energy Secretary demanded in 1975 and 1978 respectively that the company should be 100 per cent publicly owned, with full operational control by Whitehall. Those demands were officially dismissed as having been made without cabinet authority. Even so, the revelations in the Bingham report on Rhodesian sanctions, and other examples of the company's wilfulness, led senior ministers to put in hand a review of relations between Whitehall and BP.[28]. The general election of 1979 supervened, and the new Conservative government not unexpectedly opted for the alternative solution. As the directors themselves wanted, in due course it returned BP entirely to private hands.

3 Conclusions

As the above account has made clear, British governments over the years have never pursued a consistent line of policy towards oil as, say, those in France and Italy have done for at least some of this period. To ensure adequate supplies of oil products, until the early 1970s intervention by Whitehall tended to be sporadic and in the main prompted by fears of serious market failure. Hence Britain has experienced overall supply difficulties only during two world wars and temporarily during the Suez affair of 1956 and the OPEC crisis of 1973–74.

At the same time, those governments, regardless of party, have apparently accepted prices for oil products that were above the competitive level. The latest (1990) official investigation into petrol pinpointed a 'public attitude to the oil industry which was at best suspicious and sometimes actively hostile', one that (cf. Section 2.2 above) was evident even before the First World War.[29] The 1965 Monopolies Commission report had, as described earlier, noted the consistent striving of petrol suppliers, since the earliest days, to minimise price competition by one means or another, and that of 1979 had presented evidence of a 'complex monopoly' with, for instance, price or other discrimination against certain retailers. Not until the 1990 report was the general level of petrol prices in Britain, net of tax, found to be not unreasonable if compared with prices on the Rotterdam spot market or those in Western Europe generally. Why were earlier governments not more responsive to this disgruntlement, and why did they fail to enforce greater competitiveness?

There seems to have been no lack of political will in the face of overmighty oil giants. The bold acquisition of the Anglo-Persian shares in 1914 carried the risk of seriously alienating Shell and Standard Oil which then provided the bulk of Britain's oil, but the government accepted that risk. While a Conservative-dominated coalition government allowed the wartime Profiteering Act to lapse in 1921, despite the strictures of the two recent reports on petrol, it successfully gambled on its draconian deflationary measures driving down the excessive petrol prices. Although, as already shown, in 1955 certain departments' lobbying prevented petrol from being investigated by the Monopolies Commission, referral was made in 1960 by the Conservatives, at a time of growing worries about Britain's competitiveness; the commission reported to a Labour government in 1965. Most of the tough measures taken against North Sea licensees after 1974 would have had to be introduced even had Labour not won power that year. Such examples of government decisiveness undoubtedly conceal much inter-departmental and ministerial infighting, but do not suggest that the oil companies had things all their own way.

In short, governments were seeking to come to terms with the dilemma mentioned at the outset, that they had the duty of seeing that oil supplies were forthcoming and prices reasonable, but must at the same time ensure that British-owned companies or concessions were not allowed to slip into foreign control. In the early 1900s government action saved for Britain the crucial Burmese and Persian oil resources, and later in the century the British oil industry, comprising smaller companies as well as BP and Shell, flourished on an international scale and thereby materially assisted the country's balance of payments. While Jones and Kirby, in Chapter 1 above, have stressed the low productivity, slow growth and relative decline of much of Britain's industry, there *were* exceptions, and the successes in oil, as in (say) chemicals and pharmaceuticals, suggest that the causes of failure may be more complex than many scholars have so far realised. Since this question is of more than purely historical interest in Britain's current plight, perhaps the present chapter may offer a new direction for future research.

Notes

1 My thanks are due to the other participants at the Reading-Lancaster conference for their helpful comments, and to the staff of the Institute of Petroleum Library for giving me access to its collection of press cuttings from the 1960s onwards.

2 The development of petrol retailing in Britain is outlined in chapter 2 of the Monopolies Commission report on the *Supply of Petrol to Retailers in the United Kingdom*, London, British Parliamentary Papers (BPP) 1964–5, XIX, pp. 471–667. Figures in this paragraph are taken from Geoffrey Jones, *The State and the Emergence of the British Oil Industry*, London, 1981, p. 32 and the *Petroleum Times* jubilee number, 17 June 1949, p. 428.

3 Jones, *The State*, p. 9; *idem*, 'The oil fuel market in Britain 1900–1914: a lost cause revisited,' *Business History*, XX, 1978, pp. 131–52.

4 T. A. B. Corley, *A History of the Burmah Oil Company I. 1886–1924*, London, 1983, the CID memo. being mentioned on p. 90. For subsequent history see also R. W. Ferrier, *The History of the British Petroleum Company I. The Developing Years 1901–32*, Cambridge, 1982.

5 Oil control in the First World War has not been well researched, but see Marion Kent, *Oil and Empire: British Policy and Mesopotamian Oil 1901–20*, London, 1976, Appendix VI. Cowdray's career is well described in David J. Jeremy (ed.), *Dictionary of Business Biography*, IV, London, 1985 (s.v. Weetman D. Pearson), pp. 582–94; see also Jonathan C. Brown, 'Domestic politics and foreign investment: British development of Mexican petroleum 1889–1911', *Business History Review*, LXI, 1987, pp. 387–416.

6 Sub-Committee of Standing Committee on Investigation of Prices, *Reports on Motor Fuel*, London, BPP 1920, XXXIII, pp. 573ff. (Cmd. 597) and 1921, XVI, pp. 793ff. (Cmd. 1119).

7 D. J. Payton-Smith, *Oil: A Study of War-Time Policy and Administration*, Civil History of the Second World War, London, 1971, p. 40.

8 Some of the oil company measures are discussed in Corley, *A History of the Burmah Oil Company II. 1924–66*, London, 1988, chapter I. The only substantive source of information on Achnacarry is US Senate Subcommittee on Small Business, *The International Petroleum Cartel*, Washington, DC, 1952. The forthcoming second volume of Ferrier's BP history should yield further information.

9 Pre-war policy is outlined in Payton-Smith, *Oil*, and in Corley, *Burmah Oil II*.

10 Payton-Smith, *Oil* is essential reading here. See also Louis Turner, *Oil Companies in the International System*, London, 1978, chapter 3. For post-war measures see Roger Bullen and M. E. Pelly (eds), *Documents on British Policy Overseas I*, Volume IV 1945–6, London, 1987, p. 266 and 'The new oil refineries', *Economist*, 10 July 1948, pp. 69–70. For Anglo-US relations in 1950 see Bullen and Pelly, *Documents on British Policy Overseas II*, Volume II 1950, London, 1987.

11 Volume II of Ferrier's BP history is awaited here also. Kenneth O. Morgan, *Labour in Power 1945--1951*, Oxford, 1984, pp. 464ff. is essential reading on the political side. For post-1951 policy see Robert Rhodes James, *Anthony Eden*, London, 1986, pp. 346, 360; the 1953 coup is mentioned in Christopher Andrew, *Secret Service*, London, 1985, p. 494.

12 Turner, *Oil Companies*, pp. 49–50.

13 Monopolies Commission, *Petrol*, 1964–65, 'Note of dissent by T. Barna', p. 177. Cf. Political and Economic Planning, *A Fuel Policy for Britain*, London, 1966, p. 95. Reasons for high UK prices are given in M. A. Adelman, *The World Petroleum Market*, Washington, DC, 1972, pp. 225, 240.

14 Monopolies Commission, *Petrol*, 1964–65.

15 Rhodes James, *Eden*, p. 460; Turner, *Oil Companies*, pp. 101–2; for OEEC see *ibid.*, p. 52.

16 This distinction was made to the author, on his joining the Ministry of National Insurance, by the Establishments Officer.

17 There is an immense literature on every aspect of the North Sea bonanza. Recommended here are Guy Arnold, *Britain's Oil*, London, 1978, as a general

account, and Michael Jenkin, *British Industry and the North Sea*, London, 1981, helpful on government North Sea policy.

18 House of Commons Papers 1972–73, *First Report from Committee of Public Accounts*, 'North Sea oil and gas', London, 1973.

19 W. N. Medlicott, Douglas Dakin and M. E. Lambert (eds), *Documents on British Foreign Policy 1919–39*, Second Series XV, London, 1976, pp. 332ff., 360ff. and 699–700.

20 Foreign and Commonwealth Office, *Report on the Supply of Petroleum Products to Rhodesia*, by T. H. Bingham and S. M. Gray, London, 1978; Martin Bailey, *Oilgate: The Sanctions Scandal*, London, 1979, and Clive Ponting, *Breach of Promise: Labour in Power 1964–1970*, London, 1989, quoting from confidential US government documents.

21 Ian Skeet, *Opec: Twenty-Five Years of Prices and Politics*, Cambridge, 1988, gives a sober account of sensational events.

22 'The bonanza that eluded Burmah's grasp', *Business Week*, III, April 1978. Cf. the contacts between the British-owned Ultramar and Middle East countries from 1973 onwards, *Golden Adventure: The First 50 Years of Ultramar*, London, 1975, pp. 206–8.

23 *Parliamentary Debates (PD)*, House of Commons, 27 March 1972, col. 34.

24 Louis Turner, 'The European community: politics of the energy crisis', *International Affairs*, L, 1974, pp. 405–6, and Mira Wilkins, 'The oil companies in perspective', *Daedalus*, CIV, 1975, esp. pp. 169ff.

25 For meetings with Heath see Anthony Sampson, *The Seven Sisters*, London, 1975 p. 263; Robert Stobaugh, 'The oil companies in the crisis', *Daedalus*, CIV, 1975, p. 189; Harry van Seumeren, *Gerrit A. Wagner: Een Loopbaan by de Koninklye*, Utrecht, 1989, pp. 116–19, and 'The appalling candour of Gerald [sic] Wagner', *Economist*, 8 December 1973, p. 83.

26 Monopolies and Mergers Commission, *Petrol: A Report on the Supply of Petrol in the UK by Wholesale*, London, Cmnd. 7433, 1978–79.

27 *Select Committee on Nationalised Industries, Sub-Committee B, 1977–78*, 'British National Oil Corporation', Minutes of Evidence, 15 February 1978 (Lord Kearton), London, 1978, p. 41.

28 *Daily Telegraph*, 11 February 1975; *The Times*, 18 October 1978; *The Guardian*, 18 October 1978.

29 Monopolies Commission, *The Supply of Petrol: A Report on the Supply in the United Kingdom of Petrol by Wholesale*, London, Cm. 972, February 1990, p. 309. For pre-1914 outcry see Robert Henriques, *Marcus Samuel*, London, 1960, pp. 541–51, 564.

10 *Jonathan Brown*

The State and agriculture, 1914–72

1 The growth of agriculture since 1945

'Today', wrote Lord and Lady Donaldson in 1969, 'agriculture is one of Britain's most efficient industries. It compares favourably with its foreign competitors, has a controlled growth of 3½ per cent a year, and in the last ten years its labour productivity has increased at twice the rate for industry as a whole.'[1] They were expressing the common sentiment amongst politicians, much of the farming world and the press in the years immediately before the United Kingdom joined the European Economic Community. British farmers, it was held, had nothing to fear from the Common Market, for their industry was as efficient as agriculture anywhere else in Europe.

There was much to support such a view, for British agriculture had grown in many ways, had indeed been transformed since the end of the Second World War. The output of British farming had increased considerably in those years. The production of wheat had more than doubled: 4.4m tons in 1970–72 compared with 1.9m tons in 1945–47. The acreage devoted to the crop had changed more slightly, increasing from 2.2m acres to 2.6m acres in the same period. Barley, on an acreage increased from 2.1m to 5.4m acres, was producing 8.6m tons in 1970–72 compared with 1.9m tons in 1945–47. Production of sugar beet had likewise increased by about three-quarters, and milk by more than a third. Yields had risen: wheat from 19 cwt an acre to 32, barley from 19 to 30, potatoes from 7 tons an acre to 10, and sugar beet from 11.5 to 14.5 tons per acre, all between 1944–46 and 1965–67. Yields of milk per dairy cow rose from 691 gallons in 1955 to 902 gallons in 1975.

Farming had become more capital intensive, with a mechanisation that was rapid and thorough. Horses had disappeared, there were twice as many tractors on British farms in 1971 as there had been in 1946, more than three times as many combine harvesters, and a similar growth of machine milking. The employment of labour had fallen by more than half. Full-time male employees in agriculture in Great Britain declined in number from 580,783 in 1948 to 280,598 in 1968.[2] That, and the investment in machinery, buildings and other improvements, had brought about a marked increase in labour productivity. The statement of the Donaldsons was based on the work of Sharp and Capstick, the fullest of a number of studies of productivity in agriculture during the 1960s. They showed labour productivity in agriculture to have risen at 5.1 per cent a year between 1954 and 1964, while for the British economy as a whole the rate of growth was 2.5 per cent per year.[3] One result of the increases in production was that more of the nation's food was being produced at home. By the late 1960s just over half of all food requirements, and about two-thirds of those temperate-zone foods suitable for the land of this country was being supplied by British farmers. After a very rapid increase during and just after the Second World War, the trend had been gradually rising. A substantial share of all the major types of 'native' food, except butter, was now produced at home: 41 per cent of wheat supplies, 90 per cent for barley, 73 per cent for beef and veal. Compared with the years between the wars, when no more than 15 per cent of wheat supplies were grown in this country, the change was striking.[4]

All agreed that the support provided by the government was a major contributor to the extent and nature of the changes in farming. By the early 1970s a most impressive structure of guaranteed prices, subsidies, tariffs and quotas had been built up to help the farmer make a living. The foundation of it was the system of guaranteed prices, with deficiency payments from the Exchequer to make up the difference between guaranteed and market prices. The annual farm price review which determined what the guaranteed prices should be from year to year was the focal point in the operation of this system. Besides the guaranteed prices there were numbers of grants. After some considerable increase during the 1960s there were more than twenty grants available in the early 1970s. They included the calf subsidy, the hill cow subsidy, the hill sheep subsidy, and the farm capital grants, which embraced a number of grants for improvements to buildings, reclamation of land, purchase of machinery, and other investments. Added to these were the subsidised advisory services, while on overseas trade there were tariffs on some products, such as barley, and a number of agreements with individual suppliers of bacon and cereals intended to give some protection to the British farmer by preventing surpluses of produce being built up on the British market.

The cost to the Exchequer of the price guarantees and grants was in 1971–72 £338 million, considerably greater than the £206 million spent in 1955–56. That was not quite the highest total, as £342 million had been achieved in 1961–62, but after restraint in the early 1960s, government expenditure on agriculture had embarked on an upward trend from 1967 onwards. For the farmers these sums were important. Calculations by the National Farmers Union in the mid-1960s reckoned that subsidies accounted for about four-fifths of most farmers' earnings, while for numbers of small farmers in uplands the whole of their income effectively came from the government.[5]

2 Agricultural policy, 1914–72

The system of price support and subsidies in operation in 1972 had been established by the Agriculture Act, 1947. Part I of that Act had provided that the Minister of Agriculture should produce an annual review of the farming industry, and that he was to undertake this review in consultation with the representatives of the farmers, which in practice meant the leaders of the National Farmers Unions of England, Scotland and Northern Ireland. The review was to consider projected trends in production and to set the guarantees of prices for the following year. This procedure was still followed in 1972, but much had happened in the course of the twenty-five years the Act had been in force. The emphasis in the government's thinking had shifted from concern over possible food shortages, to the prevention of surpluses, and to the need to reduce imports, while the annual reviews often reflected immediate, short-term considerations in the management of the economy. There had been an increase in the number of grants, and the government's disbursements in this form had grown as a proportion of the total spending on agricultural support until by the late 1960s they were outstripping the sums spent on price guarantees. The number of overseas trading agreements also increased during the 1960s.

The Agriculture Act of 1947 had risen out of the experiences of the Second World War. Then agriculture, in common with other industries, had been taken into direct government control, with all marketing handled by the Ministry of Food, and the Ministry of Agriculture setting the targets for production, to be administered through the War Agricultural Executive Committees. From 1943 the Ministry's determination of the production targets took the form of an informal precursor of the annual farm price review, with some consultation with the farmers' leaders. The main concern of the government during the war was to increase production in agriculture, through the massive campaign to plough up grassland for crops of cereals and potatoes. There was much to be done, for agriculture in the 1930s had been at a particularly low ebb. Arable production had declined, there was widespread dereliction and arrears of maintenance, such that the first tasks for many of the county war agricultural executive committees included the reclamation of thousands of acres of abandoned or barely farmed land. As many as 20,000 acres were taken into the Essex committee's hands to be brought up to standard for cultivation, the Norfolk committee undertook the restoration of 6,000 acres in Feltwell Fen, while other committees had similar projects in hand during 1940 and 1941. Against this background the thoughts of officials at the Ministry of Agriculture and of policy-makers in the political parties turned towards establishing ways of maintaining British farming in a reasonable state of productiveness.[6] These efforts prepared the way for the Act of 1947.

The government had further experience from the 1930s upon which to draw. The Wall Street Crash of 1929 and the depression that followed brought about a reassessment of economic policies, in which agriculture had a place. Even so, the changes took a little time to materialise. While European countries hastened to raise tariffs to protect their farmers, British markets remained open. It was not until another financial crisis in 1931 forced Britain off the gold standard, and ushered in the national government prepared to abandon free trade that measures of any substance in support of agriculture were introduced.

Tariffs were imposed on the imports of a number of agricultural products. The first were anti-dumping duties placed mainly on horticultural produce, especially soft fruit and potatoes. There had been considerable outcry about the activities of some overseas suppliers in recent years in dumping produce on the British market and the new government acted quickly to take account of that through the Horticultural Products (Emergency Duties) Act of November 1931. A general tariff covering the greater part of the nation's overseas trade was introduced at the rate of 10 per cent *ad valorem* by the Import Duties Act, passed in February 1932. Although dairy produce and barley were covered by these duties, several of the major products of agriculture were omitted, including wheat, meat, livestock and wool. They were brought into the protective system following a series of complex negotiations with the leading overseas suppliers of food. A duty of 2s per quarter was imposed on wheat following agreements made at the Imperial Economic Conference held at Ottawa in July 1932. Individual agreements with Australia, New Zealand and Argentina over beef and mutton, and with Denmark over pigs and bacon resulted in the setting of tariffs and quotas for meat and livestock.[7] The appearance of comprehensive protection was illusory, however, for so many concessions were made to the various interests — imperial preference to the Dominions, reduced tariffs and guaranteed quotas to other countries — that the British farmer remained almost completely exposed to foreign competition. The most substantial change was that the Dominions accounted for a greater proportion of Britain's total imports of food. It hardly amounted to the full-scale protection that farmers wanted.

Farmers fared better from the domestic measures of support. These fell into two categories. One was the effort to strengthen the farmers' marketing position. The Agricultural Marketing Acts of 1931 and 1933 gave statutory rights of monopoly traders to producer-controlled marketing boards, provided two-thirds of the producers affected by a proposed scheme voted in its favour. Five marketing boards had been set up under the terms of the Acts by 1939, the boards for potatoes, hops, pigs, bacon and milk. The other form of support was subsidy. The Wheat Act of 1932 offered arable farmers a guaranteed 'standard price' of 10s per cwt, which would be made up to farmers when the market price fell short. It was to be a self-financing scheme, by which it was meant that the Treasury was not intending to foot the bills. Instead, a levy was to be imposed on all wheat flour delivered from the mills. As imported wheat accounted for about 85 per cent of all flour milled in this country the levy could be regarded as a disguised tariff.

It was not excessively protective, however. Imports of wheat continued to flow into Britain, mainly because the millers relied on supplies of the hard wheats grown overseas for their bread flour.[8] The Act had immediate effect on farming in Britain. Ninety-three thousand acres were added to the area under wheat in Great Britain in 1932, 7.5 per cent more than the total in 1931. The total for 1933, after a full year of the Act's operation, showed an increase of 32 per cent over the acreage in 1931. Guaranteed prices were extended to cover barley and oats by the Agriculture Act of 1937, as the government sought to reinvigorate arable farming before war came. There had been plans for a similar levy-subsidy for beef, but they had foundered on the protracted negotiations with Australia and Argentina over tariffs and quotas. Instead a basic subsidy of 5s per live hundredweight for cattle was adopted.

The subsidies introduced in the 1930s in their turn had their antecedents, this time dating from the First World War and its aftermath. The exigencies of war made urgent an increased production of food at home in order to save valuable shipping space. In the early years of the war the government had expected this increase to occur almost automatically as higher market prices induced farmers to sow more wheat. As well as that, farmers were expected to remember their sense of national duty, to which government ministers regularly appealed. The ministers were to a large extent disappointed. The higher prices had offered insufficient inducement for many farmers to make major changes to their patterns of cropping or stock management. Costs, after all, also were rising, a number of things, such as fertilizers and implements, were in short supply, and the war was not expected to be long enough to justify changing a successfully established system of farming. It was more simple for farmers to cash in the higher prices to make their first big profits in a lifetime.[9]

Faced with a shortfall on home production and with increasing losses of shipping, the government of Lloyd George stepped in to take control of agriculture. In 1917 a ploughing up campaign was launched to increase production of wheat and potatoes, powers were taken under the Defence of the Realm Act to give the newly-constituted war agricultural executive committees for each county the authority to make ploughing up orders, and to direct the supply of labour and implements. A new Corn Production Act was passed to give separate statutory authority for these executive powers. The Act also laid down the principle of minimum wages for agricultural labourers, to be settled locally by county wages boards at a rate not less than 25s per week for full-time adult males. In contrast, as an encouragement to farmers to grow more corn, the Corn Production Act prescribed guaranteed minimum prices for wheat and oats, with provision for deficiency payments to maintain the guarantees. Here, then, was the first appearance of guaranteed farm prices, and it proved to be a purely nominal one. The guaranteed prices for 1917, written into the Act, 60s per quarter for wheat and 38s 6d per quarter for oats, were considerably less than the 78s a quarter for wheat and 55s a quarter for oats which the Ministry of Food was paying in the spring of 1917, and this gap between guaranteed and actual prices remained throughout the war and immediately afterwards.[10]

The Corn Production Act extended guaranteed prices for six years. Before that term was fulfilled the Act was superseded by the Agriculture Act of 1920. Following the recommendation of the Royal Commission on Agriculture appointed in 1919 the new Act made the guaranteed prices permanent, subject to termination at four years' notice. The minimum wages and the provisions for county committees to promote efficient husbandry were also carried into this Act. It was one of the more short-lived pieces of legislation, for Part I of the Act dealing with the guaranteed prices and minimum wages was repealed in 1921 when the Treasury realised how much it was likely to cost to support the guaranteed prices once the bumper crops of cereals around the world reached this country. As the repeal Bill was going before Parliament the chairman of the West Riding branch of the National Farmers Union was declaring that the government's action was a 'betrayal' of the farmers, while proposing a resolution couched in similar terms. Almost all farmers then agreed that the government had acted with extreme perfidy in breaking so abruptly the terms set out in its own Act of the previous year. There was far less unanimity about the merits of the repeal itself. For various reasons farmers were not wholly displeased to see the Agricultural Act go. They had hated the lingering wartime controls; they hated the controlled prices, and dealing with the Wheat Commission; they detested minimum wages and the wages boards; they disliked the busybodies of the county agricultural committees, and not a few farmers thought that Part II of the Act should be repealed as well. In short farmers were desperate to be free of government controls, and were quite prepared to see guaranteed prices go as well. So it was that there were farmers at the meeting of the West Yorkshire Farmers Union who spoke in favour of the repeal. There were other branches of the farmers' union and chambers of agriculture that passed resolutions in favour of the repeal. The *Farmer and Stockbreeder's* leader writer declared that 'Part I of the Act is not a farmer's measure'. These views generally prevailed in the National Farmers Union, and were conveyed through its leadership to the government.[11]

The farmers' relief at the return of free markets was soon tempered as prices fell rapidly. By the end of the year the *Farmer and Stockbreeder* had already changed its tune. Its review of 1921 was that 'all went well until the Government repudiated the system of guarantees', with the result that the farmers' confidence 'was sadly shaken'. Before long the 'Great Betrayal' was entering farming fable, and came to acquire enhanced significance as a failing levelled against governments, which helped to ensure that agriculture would not leave the political scene.

The 1920s then became unsettled years in the establishment of policy. Bonar Law could declare in 1922 that agriculture must live on an economic basis without state subsidy or protection; and the farmers' leaders could declare that was precisely what agriculture wanted. Yet underneath there was agitation growing for a revival of government support, and farmers in the eastern counties were complaining both about the government and their union's policy. Stanley Baldwin took up a recommendation of the Agricultural Tribunal of Investigation which reported in 1923, and proposed a subsidy of 20s per acre on arable land. The farmers' leaders rejected this on the grounds that the bluntness of the instrument offered no real guarantees to farmers: they were better off finding their way in the free market. The rebuff directed the government to other measures, the subsidy to the sugar beet industry, and the encouragement to efficiency through reform of agricultural credit. The negotiations and investigations of the 1920s did mean, however, that when price guarantees returned in the 1930s, the step was relatively easy.[12]

3 Continuity in British agricultural policy

A constant thread runs through British agricultural policy from the First World War to the time of joining the Common Market. Despite all the vicissitudes and the experiments, there are common elements throughout the years. First among these is the continued employment of guaranteed prices with deficiency payments to make up the farmers' incomes. The methods of calculating the deficiency payments may have varied, but the Corn Production Act, the Wheat Act and the Agriculture Act of 1947 all shared these payments as a principal means of supporting agriculture. In comparison there was little resort to overseas trade measures in support of agriculture. The protectionist government of the 1930s introduced tariffs and quotas in number, but so beset them with imperial preference and concessions to individual trading partners that they had little practical value. After the Second World War imperial preference was less of an issue, but duties and quotas had a minor role in agricultural policy. In this Britain diverged markedly from the policy pursued by almost all the other countries of Europe, America and the British Dominions, for whom import duties, quotas and other trading measures formed a major part of agricultural protection.

Caution with government expenditure featured strongly in the direction of agricultural policy. Concern that expenditure on price guarantees might become excessive — the estimates were then £12 million for wheat and £17 million for oats — had been the major determinant in the government's decision to repeal the Agriculture Act in 1921. When guarantees did return with the Wheat Act, the levy-subsidy system was devised to ensure that payments from the Exchequer were kept to a minimum. After the Second World War, once the pressure to increase total production had passed, the government made considerable efforts to contain public expenditure on farming. 'Standard quantities' were laid down for milk from 1954 onwards to place a limit to the amount eligible for subsidy. Changes to guaranteed prices often did not match changes in costs. The government stated that it expected shortfalls to be met by increased efficiency on the part of farmers. This was a contentious issue between farmers and government, as total subventions in the farm price reviews were often well below the increases in costs: by as much as £30 million in 1958. Even with a more relaxed attitude towards public spending under Labour governments the Treasury maintained tight control, and could be accused of trying to get agricultural expansion 'on the cheap'.[13]

The strong element of continuity running through British agricultural policy owed much to the fact that most of the basic problems and issues were constant. They were present in 1917, in 1947 and still to a considerable extent in 1972. Many of these issues indeed went back far beyond the First World War; to the 1870s for example, when the influx of wheat from the prairies brought into question the farmers' ability to live with greatly reduced grain prices.

One issue in particular had been fought, and lost by the agricultural interest, as long ago as the 1840s. A commitment to low prices for food became firmly established in the British political soul with the arguments over the Corn Laws and their repeal, and it remained until finally squeezed out during the 1970s. The policy of free trade which that implied in the nineteenth century was the fundamental reason why the depression in agriculture of the 1870s–1890s elicited no deep response from government. Two Royal Commissions investigated the state of agriculture, but protective tariffs could hardly be considered; subsidies from the Exchequer had not yet become a serious possibility. Practical results, therefore, were limited to some derating of agriculture, legislation on agricultural tenancies, and the creation of a government department, the Board of Agriculture, in 1889.

When governments did turn seriously to support agriculture they remained firm in their commitment to keeping food prices low. The Wheat Act was designed to ensure that there would not be 'taxation on bread'. The whole structure of agricultural support after the Second World War was proclaimed as the cheap food policy. Government statements regularly extolled its virtues. A White Paper published in 1960, for example, was self-congratulatory of the fact that 'the retail food price index is now actually lower than it was two years ago — which in turn helps to keep down prices generally and to maintain the competitiveness of our industrial exports'. As a result, the White Paper noted, 'the housewife is able to buy food at prices lower than in most other industrial countries'.[14] The logical outcome of the pursuit of low food prices was that domestic subsidy had to take precedence over tariffs as a means of supporting agriculture because the nation was far from self-sufficient in food. The concessions made in the 1930s over imperial preference and other trade agreements derived in part from this, while in the 1950s and 1960s it was held to be one of the chief virtues of the system of guaranteed prices that markets were left free to hold retail prices down.

By the end of the nineteenth century agriculture had become a small part of an industrial economy, and the primacy of industry in economic affairs became one of the recurring issues in the formation of agricultural policy. This was most prominent, perhaps, in the 1930s, when the prospects for full protection for British agriculture were circumscribed by the need to promote industrial recovery. Duties on imported food may have discouraged the exporting countries from buying British manufactured goods; and, by increasing the price of imported food, they could have made British industry uncompetitive in international trade. Thus it was that so many concessions on tariffs were made, in order to protect industry's exports.[15] Similar arguments held after 1945: they are clearly expressed in the White Paper of 1960, that low prices for food stimulated the competitiveness of industry. These considerations, again, led the government away from import controls to guaranteed prices as the main instrument of policy.

The needs of industry could have been met as well by continuing with free trade and not giving guarantees or subsidies. That would have kept the prices of imports and therefore of all food low as successfully as any of the policies of the 1930s–1960s, and it would have allowed agriculture naturally to find its place in the economy. It remains to be considered, then, why governments should have accepted the need for a policy of supporting agriculture.

The arguments contained a mixture of economic and other social and political reasons. Prominent among the non-economic arguments was that a greater proportion of the nation's food should be grown at home to provide security in time of war. It was a favourite amongst those arguing for protection during the agricultural depression of the late nineteenth century. When war came in 1914 this argument was repeated with greater force. Daniel Hall, who was to become permanent secretary at the Board of Agriculture and Fisheries in the later years of the war, wrote in 1916 a book, the first chapter of which considered Britain's dependence on overseas supplies of food and the difficulties this was causing in the present war. He compared that with Germany which had been well provisioned from home agriculture as a result of the country's 'prevision of the conditions that war would create and by the preparation of the whole fabric of the community for the shock, in which preparedness the position of agriculture and the questions of food supplies have been matters of prime importance'.[16]

Another part of Hall's advocacy of a policy to promote agriculture was that it would 'increase the agricultural population as a specially valuable element in the community'. This had been a common argument in favour of agricultural protection since the late nineteenth century. A number of studies had drawn attention to the extent of rural depopulation, and had fuelled the argument that the benefits of fresh air and country living gave strong moral character to the population, and improved the quality of its infantrymen. Sentiment in favour of government action to safeguard the life of the countryside has remained strong in all political parties. It has taken a variety of expressions. There have been rural revivalists and back to the land campaigners who hoped to see a nation of sturdy yeomen. There has been concern at the spread of urban sprawl. The knighted Sir Daniel Hall, in evidence to the Royal Commission of Agriculture in 1919, expressed another variant of these views when he spoke of the undesirability of the countryside being 'entirely laid down to grass and to parks and to sporting estates'. To Sir Daniel that represented inefficiency and waste in having good corn land growing grass, but there was also a strong feeling that the real countryside meant cornfields.[17]

There remained a mixture of sentiments that gave general assent to the notion that agriculture somehow deserved the support of government. There were those who argued that the decline of farming had simply gone too far in the 1930s, and that farmers had been hardly treated. Here the 'great betrayal' of 1921 grew in political importance, and had not inconsiderable influence on the construction of the policy contained in the 1947 Agriculture Act, with its attempts to be fair to farmers. It had similarly been argued that it was only fair that agriculture should receive the attention of government when other industries were being supported. Winston Churchill, Chancellor of the Exchequer in 1927, noted in a memorandum that agriculture was 'still the main industry of these islands [and would] never tolerate manufacturing protection without itself being included in the tariff'.[18] Such statements were also a recognition of the strength of farming, through the National Farmers Union, as a political lobby. Conflict and disagreement between governments and union were endemic, ministers frequently had little respect for the sectional interests of the farmers' leaders, yet still had to take heed of their views. After 1947, especially, when the farmers' unions had their position in the price review procedure, they had some particular influence on the course of agricultural policy. In 1960, for example, a White Paper gave assurances of a generally expansionary trend in government support, after the leaders of the National Farmers Union had complained about the tendencies towards contraction in the operation of the Agriculture Act of 1957.[19]

Important though the arguments on the grounds of security, the virtues of rural life, or the weight of political influence could be, agricultural policy on the whole was determined by reasoning that was economic or quasi-economic in character. One powerful influence was the view that the government's support for agriculture would promote modernisation and efficiency, and would thereby enable farmers to stand up to competition from abroad. It was this view that prevailed over rural romanticism in the governments from the First World War to the Second. The goal of efficiency remained as one of the foundations of the post-war settlement in the 1947 Agriculture Act.

The Milner committee during the First World War set the ball rolling for policies to promote agricultural modernisation. Appointed to consider the production of food to meet the needs of war, the committee urged on government the policy of increasing production, especially from an increased area of arable land. To achieve this the government should promote the use of artificial fertilisers, and mechanisation, they should establish a local organisation to oversee standards of cultivation, and should raise confidence amongst farmers by guaranteeing the prices of cereals. The wider interest of Milner and his associates was declared when they stated in their final report that increased production of food at home 'will be in the permanent interest of the nation'. The country's need to re-establish its international solvency once peace was restored meant that 'the intensification of our agriculture will be even more necessary after the War than now'.[20]

During the 1920s those who took up the mantle of Lord Milner in advocating modernisation were gaining the ascendancy in the Conservative governments. Modernisation and efficiency was a theme of the White Paper on agriculture published in 1926. Having a few years since had the proposed subsidy on arable land rebuffed by the farmers' unions, and with relations between government and farmers continuing to be strained, the government refrained from embarking on a policy of direct support for agriculture. 'Agriculture', the White Paper declared, 'of all the industries in the country, is the least adapted to drastic and spectacular action on the part of the State, and its present condition is not such as to justify revolutionary methods'.[21] Instead, the government's preferred course was to promote the farmers' own efforts to modernise by facilitating credit, by education and advice, by encouraging the establishment of agricultural marketing schemes. All of these features continued into the working of agricultural policy into the 1930s, with the Agricultural Credit Act, the Agricultural Marketing Acts, and such measures as the founding of the Agricultural Research Council, and exemption of tractors from road fund tax.

The return of price support did not diminish the government's ardour for modernisation. The Wheat Act contained provision to restrict payments of guaranteed prices only to farmers who were likely to have the most suitable land, and therefore be more efficient at growing wheat. The policy of the Second World War heavily promoted mechanisation to the extent that by the war's end there were claims that British farming was the most mechanised in the world. The same spirit continued after the war, most explicitly perhaps between 1953 and 1964, when Conservative governments were aiming to contain Exchequer subsidies. The White Paper of 1960 included amongst the objectives of policy the belief that agriculture, in common with other industries, 'in order to maintain or improve its competitive position must seek to increase its efficiency year by year'.[22] The election of a Labour government in 1964 changed the emphasis, but not the underlying assumption that agriculture's future lay in improved efficiency. A White Paper published in 1965 looked to structural reform, the encouragement of farmers and workers to leave the land, and a strengthening of co-operation in the industry.[23] Agriculture Acts in 1966 and 1967 sought to put these aims into practice, together with the provision of increased production grants intended to improve the competitive position of agriculture. The next strand in the policy was the encouragement of agriculture's role in saving imports, which became of increasing importance as the country's problems with its balance of payments persisted throughout the 1960s. A strong, efficient farming industry was required for this, a view which echoed back to the arguments of Lord Milner and his committee in 1915, and of Daniel Hall in his book of 1916.[24]

Efficiency was one of the watchwords of the policy laid down in the Agriculture Act, 1947. Stability was another. The government was to promote conditions whereby farmers and agricultural labourers could expect to earn a respectable living, and farmers and landowners would see a fair return on their capital. Therein was embraced the second major economic justification for agricultural support, the fact that in an industrial economy the incomes earned in agriculture were prone to be considerably less than those attainable in most other occupations. There were fundamental reasons for that in the fact that the elasticity of demand for food is low, while there is a tendency for agriculture to oversupply its markets. In the world of international trade in which British farmers found themselves from the late nineteenth century onwards the increases in the supply of agricultural produce were coming mainly from overseas where the costs of production were so low as to bring prices on the home market below those at which British farmers could readily compete. Farming incomes suffered badly. J. R. Bellerby's calculations for the incentive income in agriculture gave a result no more than half that for industry during the great depression of the late nineteenth century, when there was no governmental assistance.[25] The falling incomes formed the basis for arguments in favour of government support for agriculture. The social argument, already met, that life and employment should be retained in the villages stood alongside another proposition, that the free market mechanism was no longer working for agricultural produce, and therefore the state must arrange conditions for farmers to earn a fair income.

The government's policy from 1931 onwards began to address the question of incomes in farming. The emphasis then was on self-help, through the marketing boards which could even out the fluctuations in prices that create uncertainty and instability. The 1947 Agriculture Act was a more concerted attempt to provide conditions of stability through the operation of guaranteed prices, and thus to raise farmers' incomes. It worked, beyond the farmers' wildest dreams. From a position of earning little more than the average worker in industry in the 1930s, the farmers were comfortably on a par with managerial incomes in the 1950s and 1960s.[26]

4 The mixed results of government policy

The idea that agriculture is a special case, unable to pay its way without subsidy had become engrained in political and economic thinking by the late 1950s. Lord and Lady Donaldson provide a clear summary of this view. Agriculture, they wrote, 'is inherently incapable of surviving entirely through private enterprise. It is unique in its relation to the community as a whole and, with the exception of the aircraft industry, stands alone in being directly dependent for prosperity on government policy. For as far as can be seen this basic inability to thrive unaided will remain because agriculture, unlike other industries, cannot prosper mainly through growth'.[27] The basic assumption was that, because the income elasticity of demand for food was low, agriculture therefore needs support and it was on the basis of this assumption that many of the judgements of farming's success since 1945 were made. The evidence called in support of those judgements was that presented at the beginning of this chapter: the increase in productivity at a rate twice that for the economy as a whole, the substantial increases in yields, the wholesale mechanisation, the fact that the land of Britain was fully cultivated, and with an increasing scale of operations as farms were being amalgamated.

There were few who dissented from these assumptions. One who did was Professor E. F. Nash. He argued along the lines of classical economics that the low elasticity of demand for food did not so much mean that agriculture merited subsidy as an exception to the operation of market forces, but that it should be allowed to find its place in the economy at which it could meet the demand for food efficiently. He suggested further that government subsidies were not likely to promote a fully competitive, efficient agriculture because they tended to disrupt the flow of resources from agriculture to industry and commerce.[28]

For demonstration of what those arguments might mean in practice one can turn to the experience of farming during the thirty years before the First

World War, when governments adhered to free trade principles, or, indeed, to the 1920s and 1930s, during the larger part of which government subsidy was absent or limited in scope. The processes of adaptation can then be seen at work. In the earlier period there was a marked reaction from the high farming of the mid-nineteenth century which had relied heavily on technical innovation to more economical forms of agriculture that did not have the tendency towards overcapitalisation that was likely in high farming. There were notable success stories in the farming of this time, such as the development of potato growing on a large scale in the Fens, or George Baylis, who made a success of extensive arable farming on the Berkshire Downs, amassing some 12,000 acres in his enterprise by 1917. There were others in the inter-war years. Several farmers made quite a success out of dairy farming, and some on a large scale such as A. J. Hosier in Wiltshire. Arthur Rickwood, East Anglia's 'carrot king' demonstrated that specialised forms of farming could be successful.[29]

Pressures towards modernisation were certainly not absent, for, although farming might retreat from the excessive use of capital towards which high farming had often led them, they did not retreat from using capital. Thus, before the First World War many farmers retained intensive feeding of livestock; the mechanisation of the cereal harvest was made almost complete; mechanised cultivation with steam traction was one of the essentials of Baylis's profitability. After the First World War, Hosier and others were taking up milking by machine, yields of dairy cows were rising, with the result that productivity in dairying was increasing at about 2 per cent a year.[30] While there was no stampede to buy tractors in these years they were being taken up where they suited the scale and economy of the farm. It is possible to postulate that much the same would have happened after 1945. The changes in technology were there, with or without government subsidy, and complete mechanisation of arable farming, the adoption of the new strains of high-yielding cereals, fertilisers, and more productive livestock were likely to have become features of modern farming whatever happened. There would have been changes to the structure of farming; indeed some of the greatest changes in the nature and scale of farm businesses after the Second World War came in those areas, such as pigs and poultry, where the subsidies were limited.[31] Here was demonstration of Nash's view that government subsidy could retard the reallocation of resources.

The problem with that sort of natural adaptation is that it is messy. There was land left uncultivated while the Baylises, Hosiers and Rickwoods were getting themselves established. It was this dereliction that the county agricultural executive committees had to clear up when the land was needed again during the Second World War. It is unfair: for every one George Baylis there were half a dozen farmers who had to give up farming because of the losses, while landlords saw their rent rolls fall. In the British experience it also meant the transfer of agricultural resources from cereals to grassland, livestock farming and special crops such as carrots or potatoes. That was upsetting to those for whom fields of wheat represented 'real' farming.

Much of the messiness and unfairness derives from the difficulties agriculture faces in adapting at the same pace as changes in market forces. Many of agriculture's problems in the 1920s and 1930s derived from the inability of farmers to keep their costs in line with the prices for their produce. The process of agricultural transformation without the government's intervention, therefore, tends to be an unbalanced one, and one that does not produce results quickly. When rapid increases in production were needed during the Second World War the government took farming in hand. The prospect of world shortages of food immediately after the war kept the same policy of promoting rapid increase in production in being, although under the modified regime of the new Agriculture Act. The danger of shortages receded, but the structure of the agricultural policy remained. Governments sought to adapt it in order to promote selective expansion, to restrain public expenditure, to save imports, and especially to develop a competitive agricultural industry.

All the governments liked to claim that they had produced an efficient and competitive agriculture, but the question then remains, competitive with what? If British agriculture was intended to be competitive with the farming of other nations then that was becoming an illusion. The agriculture of almost every nation was subsidised in some way, and international markets so manipulated that the home farmers were quite effectively shielded. Even accepting that, detailed investigation suggested that British farming was not the most efficient amongst the industrial nations of Western Europe. T. Kempinski's analysis of the financial and physical returns to farming in the Netherlands, Belgium, Denmark and the United Kingdom for the 1950s placed Britain the least of the four on almost all counts of agricultural productivity. As a result, British costs of production were higher than those of the other countries, and the incomes of farmers and labourers could only be maintained by the relatively high guaranteed prices for produce. Another decade of subsidies made little difference. C. J. Doyle's study of the position in the early 1970s showed the United Kingdom still to be outstripped by those same three countries in terms both of output per worker and output per hectare. By some reckonings other countries, such as France and West Germany, also came out well, but calculations in terms of British prices restored a balance in the United Kingdom's favour.[32]

If agriculture was intended to be competitive within the British economy, that, too, was less of an achievement than governments claimed. The much-vaunted growth in labour productivity pales beside the lack of growth in output measured against capital, so that total productivity was rising between 1948 and 1968 at 1.6 per cent per year, less than that for manufacturing industry, which was 1.8 per cent a year.[33] The guaranteed prices and capital grants were producing an agricultural industry that was returning to the high farming of a hundred years previously, with impressive technical efficiency, and little economic efficiency. This, too, was contributing to the poor international comparisons. British farming employed nearly 60 per cent more machinery and 40 per cent more buildings per £1,000 of gross output than the Netherlands, yet Dutch net output per £100 of primary inputs was nearly twice that of the United Kingdom.[34] By the end of the 1960s the scale of the capital investment in agriculture was having adverse effects on the countryside, with deep ploughing destroying ancient monuments, the loss of hedgerows, and pollution from slurry.[35]

Britain's success in keeping food prices low and in restraining public expenditure had relied on the fact that the country was a substantial importer of food. By the end of the 1960s self-sufficiency was increasing and threatening the structure of agricultural support. Even before accession to the European Economic Community British governments were having to face up to increased expenditure on agriculture and to turn to more overtly protectionist means of support.

Notes

1 J. G. S. and Frances Donaldson, *Farming in Britain Today*, 1969, p. xiv.

2 B. A. Holderness, *British Agriculture since 1945*, 1985, pp. 167–75, *Agricultural Statistics, United Kingdom.*

3 G. Sharp and C. W. Capstick, 'The place of agriculture in the national economy', *Journal of Agricultural Economics*, XVII, 1966, pp. 2–16; E. A. G. Robinson, 'The desirable level of agriculture in the British economy', in J. Ashton and S. J. Rogers, *Economic Change and Agriculture*, 1967, pp. 23–42; K. Dexter, 'Productivity in agriculture', in *Ibid.*, pp. 66–84.

4 Donaldson, *Farming in Britain Today*, p. 352; A. H. J. Baines and E. J. Angel, 'The measurement of self-sufficiency in food and agricultural products', *Economic Trends*, cxc, August 1969, pp. xxxv–xliii.

5 H. F. Marks and D. K. Britton, *A Hundred Years of British Food and Farming: A Statistical Survey*, 1989, p. 153; Donaldson, *Farming in Britain Today*, pp. 29–30.

6 P. Self and H. J. Storing, *The State and the Farmer*, 1962, pp. 20–4.

7 T. Rooth, 'Trade agreements and the evolution of British agricultural policy in the 1930s', *Agricultural History Review*, XXXIII, 1985, pp. 173–90; I. M. Drummond, *Imperial Economic Policy 1917–1939: Studies in Expansion and Protection*, 1974.

8 J. A. Mollett, 'The Wheat Act of 1932: a forerunner of modern farm price support programmes', *Agricultural History Review*, VIII, 1960, p. 31.

9 P. E. Dewey, *British Agriculture in the First World War*, 1989, pp. 79–86, 233–5.

10 Dewey, *British Agriculture*, pp. 91–103; A. R. Webber, *Government Policy and British Agriculture 1917–1939*, unpublished Ph.D. thesis, University of Kent,

1982, p. 8; Edith H. Whetham, *The Agrarian History of England and Wales, viii, 1914–1939*, 1978, pp. 94–6.

11 Edith H. Whetham, 'The Agriculture Act, 1920, and its repeal: the "great betrayal"', *Agricultural History Review*, XXII, 1974, pp. 42–9; A. F. Cooper, *British Agricultural Policy 1912–36: A Study in Conservative Politics*, 1989, pp. 47–60; *Farmer and Stockbreeder*, 13, 20 June 1921.

12 Cooper, *British Agricultural Policy 1912–36* pp. 65–9; Webber, *Government Policy and British Agriculture 1917–1939*, pp. 38–48; *Farmer and Stockbreeder*, 26 December 1921, 13, 20 August 1923.

13 Whetham, 'The Agriculture Act, 1920, and its repeal', pp. 42–9; Asher Winegarton, 'British agriculture and the 1947 Agriculture Act', *Journal of the Royal Agricultural Society of England*, CXXXIX, 1978, pp. 78–82; C. Selly, *Ill Fares the Land*, 1972, pp. 45–53.

14 Mollett, 'The Wheat Act of 1932', pp. 26–7; Selly, *Ill Fares the Land*, pp. 19–21.

15 Rooth, 'Trade agreements and the evolution of British agricultural policy in the 1930s', pp. 177–90.

16 A. D. Hall, *Agriculture After the War*, 1916, pp. 1–17.

17 Hall, *Agriculture After the War*, p. 127; Cooper, *British Agricultural Policy 1912–36*, pp. 69–71; Royal Commission on Agriculture, 1919, *Minutes of Evidence*, Q. 13.

18 Quoted in R. D. Herzog, *The Conservative Party and Protectionist Politics 1918–1932*, unpublished Ph.D. thesis, University of Sheffield, 1984, p. 125.

19 Cooper, *British Agricultural Policy 1912–36*, pp. 52–3, 64–9, 119; Self and Storing, *The State and the Farmer*, pp. 79–80; *Farmer's Weekly*, 23 December 1960.

20 Committee of the Board of Agriculture on the Production of Food, Interim Report; Final Report. Cd. 8048, Cd. 8095, 1915; E. A. Attwood, 'The origins of state support for British agriculture', *Manchester School*, XXXI, 1963, pp. 138–41.

21 *Agricultural Policy*, Cmd. 2581, 1926.

22 *Agriculture*, Cmnd. 1249, 1960.

23 *The Development of Agriculture*, Cmnd. 2738, 1965.

24 Hall, *Agriculture After the War*, 1916, pp. 96–100, 127.

25 J. R. Bellerby, *Agriculture and Industry Relative Income*, 1956, pp. 62–73.

26 J. K. Bowers and P. Cheshire, *Agriculture, the Countryside and Land Use*, 1983, pp. 79–87.

27 Donaldson, *Farming in Britain Today*, p. 25.

28 E. F. Nash, *Agricultural Policy in Britain*, 1965, pp. 14–18.

29 C. S. Orwin, *Progress in English Farming Systems, III, A Specialist in Arable Farming*, 1930: A. J. and F. H. Hosier, *Hosier's Farming System*, 1951, pp. 45–9. *Farmer and Stockbreeder*, 24 February 1930.

30 *Agricultural Statistics*, 1934, pp. 44–5; A. W. Ashby, 'The milk marketing scheme', *Agricultural Progress*, XII, 1935, p. 1.

31 I. R. Bowler, *Government and Agriculture*, 1979, p. 64.

32 Centre for Agricultural Strategy, Report No 7, *The Efficiency of British Agriculture*, 1980, pp. 44–64; Bowler, *Government and Agriculture*, pp. 27–9; T. Kempinski, *Income and Efficiency in Agriculture: A Comparison between Belgium, Denmark, the Netherlands and the United Kingdom*, University of Manchester, Department of Agricultural Economics, Bulletin no. 104, 1964; C. J. Doyle, 'A

comparative study of agricultural productivity in the UK and Europe', *Journal of Agricultural Economics*, XXX, 1979, pp. 264–7.

33 J. K. Bowers, 'Economic efficiency in agriculture' in Open University, *Agriculture*, 1972, pp. 93–100; J. K. Bowers, 'British agricultural policy since the Second World War', *Agricultural History Review*, XXXIII, 1985, p. 74.

34 Doyle, 'A comparative study of agricultural productivity', pp. 265–7, 270.

35 Bowers and Cheshire, *Agriculture, the Countryside and Land Use*, p. 24; Selly, *Ill Fares the Land*, pp. 85–8.

Index